TEXTBOOK OF BIOSYSTEMATICS

THEORY AND PRACTICALS

T. Pullaiah

2024

Regency Publications

A Division of

Astral International Pvt. Ltd.

New Delhi – 110 002

First Publishes, 2013

Reprinted, 2024

ISBN: 978-81-89233-81-5 (HB)

Published by : **Regency Publications**
A Division of
Astral International Pvt. Ltd.
– ISO 9001:2015 Certified Company –
4736/23, Ansari Road, Darya Ganj
New DelhiA110 002
Ph. 011A43549197, 23278134
E-mail: info@astralint.com
Website: www.astralint.com

Printed at : Replika Press Pvt. Ltd.

Preface

Biosystematics is being taught in many biology courses including Botany, Zoology and Microbiology. Teachers expressed the non availability of text books on this subject. This book is an answer to them. This book is mainly written to cater to the needs of Botany post graduate students and researchers; it is also useful to Zoologists and Microbiologists. There must be some short comings in the book. I request the teachers and the students to bring to my notice these short comings.

I thank Dr. G. Meerabai, Department Incharge, Botany Department, Rayalaseema University for her interest in the subject and stimulating discussion. My thanks are also to Mr. K. Raja Kullai Swamy for drawing the polygraphs.

TEXTBOOK OF BIOSYSTEMATICS

THEORY AND PRACTICALS

Contents

Chapter 1
Introduction

We love and appreciate nature one way or another. Peculiar shapes, attractive colors, and curious daily activities of many life forms occurring around us, plants and animals alike, greatly stir our curiosity. We have learned to know many of these life forms through observation and experience. In order to be able to recognize these living organisms again and again, we have to apprehend and record characteristics of their appearance, habits and various other aspects, and give each a unique name. In fact, everyone of us, in this context, is a "biosystematist".

What is Biosystematics?

Biosystematics is the science through which life forms are discovered, identified, described, named, classified and catalogued, with their diversity, life histories, living habits, roles in an ecosystem, and spatial and geographical distributions recorded. In essence, it is biosystematics, the science that provides indispensable information to support many fields of research and beneficial applied programmes (Figure 1.1).

Systematics "is at the same time the most elementary and most inclusive part of biology, most elementary because organisms cannot be discussed or treated in a scientific way until some taxonomy has been achieved, and most inclusive because systematics in its various branches gathers together, utilizes, summarizes, and implements, everything that is known about organisms, whether morphological, physiological, or ecological." (Paraphrased from George Gaylord Simpson's book, "Animal Taxonomy").

Systematics can be used, as defined by Simpson (1961) as the scientific study of the kinds and diversity of organisms and if any and all relationships between them; 1. Relationship of descent, 2. Relationships of similarity, 3. Spatial or geographic, and 4. Trophic relationship.

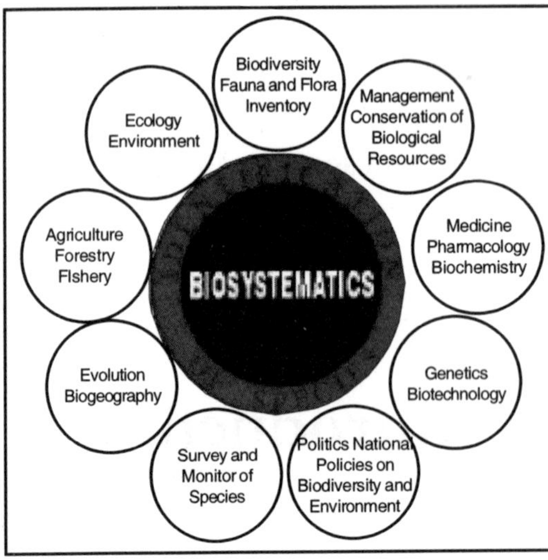

Figure 1.1: Chart of Biosystematics

Biosystematics is that part of systematic concerned with the variation and evolution of a species. It is often more concerned with the process of evolution that it is with the classification itself. Based mainly on the breeding systems several special classifications have been proposed.

Definition, History, Scope, Importance and Objectives of Biosystematics

Definition

The statistical analysis of data obtained from genetic, biochemical and other studies to assess the taxonomic relationships of organisms or populations, especially within an evolutionary framework is known as Biosystematics. The study of the variation and evolution of a population of organisms in relation to their taxonomic classification is known as Biosystematics.

The term 'Biosystematics', coined for the first time by Camp and Gilly (1943), as 'Biosystematy' to cover newer approaches in Taxonomy. Lawrence (1965) was uneasy with the above word and recoined it as Biosystematics which is accepted form scientifically. Clausen *et al.* (1945), Lawrence (1965), Valentine and Love (1958), Solbrig (1968), Grant (1971) and several others developed the subject. According to them classification and categorization of variation studies, plays an important role in which one need:

1. To delimit the natural biotic units (*e.g.*, species, genus),
2. To apply to these units a system of nomenclature adequate to the task of conveying precise information regarding their defined limits, relationship, variability and dynamic structure.

Biosystematic studies include all the relationships but emphasis on phylogenetic and phenetic relation. These relations include four types.

1. Relation of descent, also called phylogenetic relationship,
2. Relation of similarity known as phenetic relationship,
3. Spatial geographical relationship,
4. Tropicals including parasitism, commensalisms etc.

The central theme of this term is systematics with the help of cytology and genetics, help in checking the limits of taxonomic and systematic categories.

The main purpose of this branch of study is to delimit evolutionary units and devise newer taxa and check the limits of those that are already existing. The object is to work out the operative processes of speciation rather than the speciation itself.

At a particular time interval, phenotype is the final results of action, reaction and co-action between and among the factors of the environmental complex on one hand and genotypic milieau on the other. Biosystematics emphasizes on this relationship and evaluate the phenotypic expression, the result of the variation in the factors of the environmental complex and genotypic compound.

Biosystematics is a phase of botanical research that endeavors, by study of living populations, to delimit the natural biotic units, and to classify them objectively as taxa of different orders of magnitude. This necessitates use of data from the fields of ecology, genetics, cytology, morphology, phytogeography and physiology, particularly as observed from plants grown under artificial and natural conditions of environment. The systematist strives to determine objectively whether a taxon (or a population) belongs in the category of genus, species, subspecies etc. In this approach emphasis is placed on cytogenetics and cytotaxonomy supplemented by the classical approaches of morphology, ecology, and phytogeography.

The term population genetics is concerned with the gene frequencies in population. It is a combination of genetics and systematic. Camp (1961) equated it with biosystematics while Longlet (1971) to genecology. In practice, it is applied like genecology to interspecific studies in which morphological differentiation is usually limited for formal classification.

A first step in biosystematic investigations is a thorough sampling of the taxon (it may or may not be a species) and its populations and the cytological study of the chromosomes of many populations within geographic races, species, genera and so on. Differences in chromosome number, their morphology and behaviour at meiosis usually indicate genetic differences of taxonomic significance.

A second step includes the determination of the ability of the different population to hybridize, and a study of the vigor and fertility of the hybrids. This disclosed the presence or absence of breeding barriers between groups and is of taxonomic importance as indicating the natural limits of the taxa of various levels or order.

A third step studies the homologies of the chromosomes in the hybrids, as determined at meiosis. These are important indicators of the degree of genetic relationship in the material.

Information obtained from these 3 steps is compared with the data obtained from comparative morphology and geographical distribution. The resultant classification of the taxa to which it is applied (within the category of genus and taxa of lower level) has an increased objectivity over one obtained through a consideration of morphology and distribution alone but may not in all instances be an acceptable substitute for the classification resulting from the synthesis of data obtained from all sources.

History of the Biosystematists

The Biosystematists had their origin in a group of botanists initially drawn to California and its flora after the first World War and interested in a more evolutionary approach to botany. While most of the botanical (primarily taxonomy) activity in the United States at the beginning of the twentieth century centered on the Gray Herbarium at Harvard University, the balance of activity and influence began to shift westward in the early 1920s, to the University of Chicago and the West Coast. Unlike botany at Harvard, which dwelt almost exclusively on systematics (legacy of Asa Gray), botany at Chicago went in a different direction, emphasizing the newer areas of ecology and morphology, reflecting the influence and leadership of midwestern botanists such as John M. Coulter and Charles Bessey. A third centre of botany arose in the San Francisco Bay area in the 1920s and 1930s, with a focus on efforts to understand mechanisms of plant evolution using a combination of approaches that involved ecology, genetics and systematics (Smocovitis, 1994).

The most important of these early groups of plant geneticists and evolutionists who were beginning to make their way to California by the early 1920s were those located at the University of California in Berkeley and the Carnegie Institution of Washington, based at Stanford University in Palo Alto. Harvey M. Hall, Ernest B. Babcock and later G. Ledyard Stebbins, Jr. at UC Berkeley and the Carnegie team of Jens Clausen, William Hiesey, and David Keck, were instrumental in bringing together a group of diverse and interactive botanists at institutions in the Bay area devoted to understanding evolution in plants using a combination of experimental approaches. Due largely to their efforts and the close proximity of other institutions such as the California Academy of Sciences, there developed considerable interest in what were then regarded as newer and bolder, more cross-disciplinary, approaches to understanding the "new" systematics which was gaining momentum at that time (Hagen, 1984; Smocovitis, 1994, 1997). Encouraged by this growing interest in evolution and systematics, ties among local biologists were further strengthened with the creation of an informal society in the Bay area in the mid-1930s with the name of the "Biosystematists". The Biosystematists began as a very informal organization, with the group meeting on a monthly basis at one of the Bay Area institutions, usually in the evening for dinner and a discussion of current research or issues led by one of its members or invited visitors. For many years, only members and invited visitors were allowed to attend (Smocovitis, 1994). Although the exact date of its founding is still debated and membership records for the early years were not kept, the first "official" meeting of the group is thought to have been held in October 1936, at the Carnegie Institution of Washington at Stanford, with David Keck leading a discussion on plant biogeography and Wegener's then controversial theory of continental drift (Smocovitis, 1994, 1997). By the late 1930s like-minded zoologists in the Bay area, such as Richard Goldschmidt and I. Michael Lerner, a close friend of the eminent evolutionist Theodosius Dobzhansky who was one of the "architects" of the modern evolutionary synthesis (then at California Institute of Technology and a frequent visitor to the Bay Area), were drawn into the group.

By the early 1940s, the Biosystematists was one of two 'local' organizations in the United States that were to play a greater and greater role in focusing the evolutionary activity of geneticists, paleontologists, systematists and naturalists and in helping to organize a formal society for the study evolution at the national level (the other being centered at institutions in the New York City area; Columbia University and the American Museum of Natural History) (Smocovitis, 1996). The culmination of these efforts led to the formation of an organizing "Committee on Common Problems in Genetics, Paleontology, and Systematics" sponsored by the National Research Council, and eventually the founding of the Society for the Study of Evolution (SSE) in 1946 at meetings in St. Louis. Prominent among the active members of the former committee were a number of California botanists including Babcock, Ralph Chaney, Herbert Mason and Stebbins of UC Berkeley, as well as Edgar Anderson and Carl Epling of the University of California at Los Angeles. All were also active participants at meetings of the Biosystematists during this period.

Why is Biosystematics so Important?

Biosystematics knowledge of living organisms is essential for programs designed to preserve the integrity and well-being of the life-sustaining systems of the biosphere and to benefit humanity, including:

1. Comprehensive use, management, conservation and protection of the earth's biological diversity and resources;
2. Preservation of the earth's ecosystems and environments;
3. Development of human society in a sustainable manner in order to preserve the vital equilibrium of nature;
4. Discovery and identification of new food sources, genetic resources, environmental bio-indicators, biological control agents, and organisms with medicinal and other beneficial properties.

Objectives of Biosystematics

1. How different individuals of the same species acquired by physiological, morphological adaptive characters ameliorated and fixed in their genotypes in different geographical regions.
2. Based on their crossability and gene exchange among the individuals how the respective category will be established.
3. How each category is equated with the classical taxonomic – counterparts in hierarchial category.

Value of Biosystematics

1. Biosystematics provide Classifications for our millions of species. Provide natural classifications which reflect evolutionary history and is based on sound phylogenetic analyses.
2. Allow prediction of attributes of taxa not yet studied - medicines (antibiotics, etc.), biological control agents, predict ecological relationships like extinct taxa.
3. Conservation Biology - Biodiversity crisis- Massive habitat destruction- ca. 1 species extinction / 20 minutes- current extinction rates 100 to 1,000 times greater than "normal". Natural classifications allow predictions.

Role of Biosystematics in Understanding Evolution

Biosystematics, in contrast to classical taxonomy, concerns itself primarily with the process of evolution and only secondarily with its end products. The distinctive feature of biosystematics is considered to be its emphasis upon the process responsible for differences between taxa, rather than the differences themselves.

Systematics also helps us understand the process of evolution, which is information used by many other areas of biology. The microprocesses of evolution, including individual variability, population variation, reproductive isolation, modes of specieation etc., are all revealed through systematic studies. Some workers prefer to call these kinds of investigations "evolutionary biology" or even more specifically, reproductive biology, population genetics, speciation biology, etc., but they all clearly fall within systematics. The populational data are used by areas such as genetics, developmental biology and even more distant subdisciplines such as game theory. Broaderscale evolutionary phenomena, sometimes called macroevolution are also revealed through systematic studies, for example, trends in the specialization of seeds and seedlings and many other reproductive features of flowering plants. These broader insights are likewise useful for other areas of biology (*e.g.*, anatomy, morphology and developmental genetics).

Systematic studies also help reveal patterns of evolution that are useful and stimulating to other areas of biology. Patterns resulting from evolutionary process occur at all levels of organization from the local population to ordinal and class lineages hundreds of millions of years old. The ancestor-descendent and associated patterns of relationship over long periods of evolutionary time, called phylogeny and their reconstruction are of special interest to systematic. Phylogeny is important because it has much to do with the construction of classifications.

Chapter 2
Biosystematic Categories

Seeds and propagules of different plants are migrating to different geographic localities. For the sake of better adaptation each individual ameliorates itself by producing heritable physiological, morphological adaptive characters with which it will be surviving when they are brought to the experimental gardens, tested for their phenotypic expression and tested for their gene exchange through crosses. Thus each category will be established, they will be equated with the taxonomic counterparts in hierarchial category. First type of category indicates intraspecific categories among the individuals of same species or genus.

Biosystematic Categories

The major objective of biosystematic studies is to arrive at a better understanding of the natural relationships of plants, particularly those of the rank of genus and below. These categories are not intended as substitutes for the units used in classical or practical taxonomy, and they are necessarily the equivalent of these, although they may be counterparts of them. Each provides a single-word term for biosystematic situation, and in no case should the term be applied to a plant or a population unless the situation for which the term stands has been proved experimentally to exist for the particular taxon. The four most widely accepted categories of the biosystematics are, in order of ascending phyletic value, ecotype, ecospecies, coenospecies and comparium.

Ecotype

It is the basic unit in biosystematics. The term ecotype proposed first by Turesson (1922) was well defined by Gregor *et al.* (1936) as "a population distinguished by morphological and physiological characters, most frequently of a quantitative nature; interfertile with other ecotypes of the ecospecies, but prevented from freely exchanging genes by ecological barriers".

Nature and Origin of Ecotypes

Ecotypes do not originate through sporadic (fortuitous) variation preserved by chance isolation. They are on the contrary to be considered as effect of the habitat factor, products arising through sorting and controlling. The ecotype concept implies that some population fractions can be demonstrated to be adaptively related in respect of certain feature to a particular environmental factor within the species range. The basis of gene-ecology, as Harberd (1958) says is the observation that this infra-specific variability can be demonstrated to be of genetic origin and is not randomly dispersed throughout the species range, but is distributed in such a way that neighboring plants tend to resemble one another. The adaptation can be demonstrated in morphological or physiological characters.

In ecotype differentiation adaptation refers to any feature of an organism or its parts which is of definite value in allowing that organism to exist under conditions of its habitat (Daubenmire, 1959). It is recognized, however, that all the factors of the environment are interrelated and contribute to the selective forces operating on natural populations. A single genotype may be able to grow successfully under a wide range of conditions, often assuming different morphological forms. The adaptability of the race lies in its ability to grow under a wide range of habitat conditions.

Not all genotypic differences between population will be adaptive. Environmental selection of genotypes most fitted to a particular habitat, and selective elimination of unfitted genotypes giving genetically adapted will take place and the ecological product is the ecotype.

The ecotype is an experimental category, so that it's adaptive nature to be proved by experiment. The different results of environmental selection have both been referred to loosely as ecotypes, it has been suggested that the genetically adapted fixed situation be distinguished as a *genecotype.* The situation where it is the flexibility that is genetically determined and parallels the genecotype in its response to environmental conditions might usefully be termed as *phenecotype.* Under conditions of geographical isolation as on islands, so called insular races may be built up and be considered as *geo-ecotypes* which have not been formed as the result of selection.

Populations which are morphologically, physiologically and developmentally adapted to live under different environmental or habitat conditions are ecotypes.

Turesson put forward three basic propositions to explain the origin of ecotypes.

1. A species with wide ecological amplitude shows variations in morphological and physiological characters in different areas.
2. The differences are related to habitat variation.
3. The differences are not due to plastic response to change in environment but are actually due to natural selection to locally adapted populations from a pool of genetic variations available in the species.

Difference in ecotypes, therefore, result from natural or discriminating selection of biotypes by unlike habitats. Therefore, the wider the ecological amplitude of a species the more numerous are its ecotypes.

Since the ecotypes are inter-fertile they produce hybrids where the range of two ecotypes overlap. Out side this region the ecotype exist unaltered. The hybrids between ecotypes are, however, incompletely adapted to habitats of their parents, therefore their number is never large. By chromosomal doubling certain ecotypes are formed, polyploidy often leads to the formation of new ecotypes. Since polyploidy seldom exhibits same ecological tolerance as their parents, loss or addition of chromosomes segments also produce changes in genotypes leading to the formation of new ecotypes (polyploidy and aneuploidy).

Significance

Formation of ecotypes helps the species to external ecological range and spread to new areas. Cultivation of economically important plants in climatically and edaphically different places has been made possible due to evolution of new ecotypes. It is done by planting a number of ecotypes and their hybrids in an area. One or the other ecotypes shows adaptation to the new habitat, morphological variations can be marked in the species growing on varying habitats which lead to evolution.

Recognition and Classification of Ecotypes

Any feature of an organism or its parts which is of definite value in allowing that organism to exist under the conditions of its habitat, is called Ecotypic differentiation or adaptation.

The Ecotype status of population depends primarily on habitat recognition. A relation need to be established between the population hereditaty and the prevailing habitat factors. Ecotypes are therefore recognized as the product of the selective action of the whole environment, but they are defined with particular environmental characteristics. As Gregor puts it, one is forced to select that part of the heredity which is related to certain, more conspicuous features of the habitat. The factors other than these may had a part in the pattern of variation observed need to be remembered.

Three main sets of Ecotypes are normally recognized, climatic, edaphic and biotic depending on which of the environmental factors has been the most dominant. Some other kinds of ecotypes have been described.

1. Climatic ecotype (Turesson, 1930), syn: Climatype, *e.g., Leontodon autumnalis*
2. Edaphic ecotype (Ramakrishna, 1961, Gregor, 1942), *e.g., Euphorbia thymifolia,*
3. Biotic ecotype (Sinskaja, 1931), *e.g., Portulaca oleracea,*
4. Geographic ecotype, syn. Reclussion type (Turessor, 1927), geoecotype (Gregor, 1931), *e.g., Hieracium umbellatum,*
5. Climatic-edaphic ecotype, *e.g., Cenchrus ciliaris* (Pandey and Jayanth, 1970).

One must bear in mind that selection operates not on single characters but on character complexes, even though one or a few characters may found to show variation related to the chosen environmental variables. Overemphasis on single characters may, as Stebbins (1950) points out confuse rather than clarify our understanding in the variation pattern. The ecotypic fraction of a species or population in response to one selected feature of the habitat may not coincide with the subdivision in response to another which became major source of difficulty in the taxonomic treatment of geographical differentiation. Taxonomic recognition will concentrate more on this individuality than on the demonstrable ecotypic picture, and in case of ecoclinally disturbed variation, taxonomic treatment is bound to be unsatisfactory, as it will weigh the characters which contribute to the individuality of the local by differentiated population and in so doing cut across clinal patterns (Heywood, 1959). The difficulties of demonstrating variants have been considered by Bradshaw (1959) and by Wilkins (1960). It may be morphological expression of an underlying physiological difference.

Ecotype and Taxonomy

When adaptive variant or ecotype show correlated differences influenced by morphological treatment, they may be considered eligible for taxonomic consideration following normal taxonomic principles. The means by which morphologically different populations are detected provided they are on a scale considered worth recognition this can be attempted after ecotypes are found to coincide with recognized taxa for examples in *Orchis latifolia, Geranium robertianum, Silene dioica, Potentilla glandulosa* or in other cases taxa have to be erected to accommodate them. Clausen *et al.* originally considered the morphologically separate climatic ecotypes as the basis of subspecies.

Many difficulties of which the principal ones will now be considered.

1. The adaptive features of the ecotype do not lend themselves to taxonomic treatment or not correlated with features which do in several cases.
2. Most morphological differences between ecotypes are quanfitative depending on polygenic inheritance, they may require statistical treatment for their detection and it may be found that the variation within the ecotypes obscures their racial differences.
3. Genecological variation in populations may frequently occur at lower level than formal nomenclature classification can usefully do. Orthodox taxonomy at microevolutionary level has severe limitations and a problem that is seldom considered is the need for taxonomic recognition of ecotypic variation (Heywood, 1959).
4. Daubenmire (1958) states that recognition of ecotypes taxonomically for ecological purposes. He further emphasizes it is clearly useful for variants to have names, so that they can be easily referred to.
5. The presence of many species in floras are common to them will suggest a degree of similarity which may be highly misleading, if the plants belong to different ecotypes (Curtis, 1959).
6. It became very difficult to recognize the micro-taxonomic units below the subspecies level. The subdivision of the species all the way down to the forma level was practiced in Pflanzenreich monographs. The lower units are too small to occupy the serious attention of taxonomist. But many

practicing taxonomists find the need for the occasional use of all the infra specific units for dealing with the variation in well studied groups.

7. Cytological behaviour. In this case karyotypes and their behaviour are observed in different forms. The differences in cytological behaviour show existence of distinct ecotypes.

All the ecotypes so related that they are able to exchange genes freely without loss of fertility or vigour in the offspring.

Ecospecies

The ecospecies was defined first by Turesson " as a group of plants comprised of one or more ecotype, within the coenospecies, whose members are able to interchange their genes without detriment to the offspring". Related ecospecies are usually separated by incomplete genetic barriers, which, in addition to ecological barriers, are adequate to preclude free interchange of genes with any other ecospecies. When ecospecies of one coenospecies are crossed, the resultant hybrids are either partially sterile, or if fertile, they produce many weaklings in the F_2 generation (as slow growing dwarfs, individuals highly susceptible to disease against which parents enjoyed immunity, and teratological misfits). Such weaklings are unable to compete, and fail to reproduce. A few such hybrid segregates may possess sufficient vigour to survive. These may be reabsorbed by interbreeding into one or the other parental ecospecies. Related ecospecies generally inhabit different but often contiguous ecological or geographical areas, thus retaining a relative genetic purity. In general, the ecospecies approximates the conventional and conservative taxonomic species.

Coenospecies

The coenospecies is a group of plants representing one or more ecospecies " of common evolutionary origin, so far as morphological, cytological and experimental facts indicate". Coenospecies of the same comparium are separated by genetic barriers so nearly absolute that all the hybrids between them are sterile unless amphiploidy (amphidiploidy) occurs. For this reason, distinct coenospecies may exist in a single environment without genetic intermixing. Rather often the coenospecies parallel the taxonomic sections or subsections of the genus. Here ecological separation does not play part in this definition. Like that of the ecotype, ecospecies, the identification of a phyletic unit as a coenospecies must be based on genetic experiment to determine the degree of fertility.

Comparium

The comparium is the biosystematic unit that often is comparable to the genus. It is composed of one or more coenospecies that are unable to intercross. Distinct comparia are unable to intercross and complete genetic incompatibility prevails between them. There are, however numerous taxa, accepted by orthodox systematist as genera, that may contain two or more comparia (*e.g.*, in the Fabaceae). In some families the accepted and conventional genera are not equivalents of comparia or even of coenospecies (*e.g.*, some Crassulaceae, Orchidaceae, Brassicaceae).

Two separate schemes have been devised by Turesson (1922) and Danser (1929). In Turesson's system there are three ranks (Coenospecies, ecospecies, ecotype). In Danser's system the definitions of the terms are such that none of the six is synonymous with any other (Comparium, Commiscuum, Convivium). Danser's lowest term convivium covers the definition of both ecospecies and ecotype while his higher terms Comparium and Commiscuum define the higher and lower aspects of coenospecies. In addition, Turesson employed a lower term still, the ecophene, which denotes the ecological variant, purely the product of environmental modification of the phenotype. F.E. Clement used the term ecad for such variation.

Table 1.1: Comparison of the Biosystematic Categories of Turesson (1922), Danser (1929), Gilmour (1939) and Heslop-Harrison (1954)

	Turesson	*Danser*	*Gilmour, Gregor &*	*Heslop-Harrison*
Individuals occupying a particular habitat and forming an interbreeding population which differs genotypically from other such populations produce fertile hybrids	Ecotype	Convivum	Genecodeme	Group capable of hybridizing with other groups to give hybrids showing complete fertility
Individuals capable of exchanging genes freely without loss of fertility or vigour in the offspring the resultant hybrids are either partially sterile, if fertile they produce many weaklings, the weaklings are unable to compete	Ecospecies		Hologamodeme	Group capable of hybridizing with other groups to give hybrids showing some fertility
Individuals capable of hybridizing with one another, produces sterile hybrids unless amphidiploidy	Coeno-species	Comiscuum	Coenoga-modeme	Group capable of occurs hybridizing with other groups but such hybrids are sterile
Individual not capable of hybridizing with one another, complete genetic incompatibility prevails between them		Comparium	Syngam-odeme	Group incapable of hybridizing with any other such groups

Deme Terminology

Gilmour and Gregor (1939) proposed a new system of terminology designed to provide an infinitely flexible series of categories which could be used to define any group of individuals on the basis of any set of criteria. This system, the deme terminology is atleast in its original concept, non-hierarchial and it falls outside the scope of formal taxonomic categories (genus, species etc.). For that reason it avoids the use of root-words such as 'species' and 'type' which are associated with the latter. Central to the idea is the use of a natural root, deme, which implies nothing in it self except a group of related individuals of a particular taxon. The precise meaning of the terms are provided by various prefixes of which only three were originally proposed.

Topodeme: a deme occurring with in a specified geographical area.

Ecodeme: a deme occurring within a specified kind of habitat.

Gamodeme: a deme composed of individuals which interbreed in nature.

Later Gilmour and Heslop-Harrison (1954) expanded these suggestions and provided seven further basic terms.

Phenodeme: a deme differing from others phenotypically.

Plastodeme: a deme differing from others phenotypically but not genotypically.

Genodeme: a deme differing from others genotypically.

Autodeme: a deme composed of predominantly self-fertilizing (autogamous) individuals.

Endodeme: a deme composed of predominantly closely inbreeding (endogamous) but dioecious individuals.

Agamodeme: a deme composed of predominantly apomictic (non-sexually reproducing) individuals.

Clinodeme: a deme which together with other such demes forms gradual variational trend over a given area.

Cytodeme: a deme composed of individuals all with the same karyotype.

Gilmour and Heslop-Harrison (1954) also proposed some second order terms plastoecodeme, gamoecodeme, hologamodeme, coenogamodeme and synamodeme which presently are regarded as equivalent to ecophene or ecad, ecotype, ecospeices, coenospecies and comparium respectively.

Chapter 3
Plant Taxonomy: Static and Dynamic Concepts

What is Taxonomy?

The literal meaning of the term taxonomy is "lawful arrangement" or arrangement by rules (from Greek, *taxis* = arrangement; *nomous* = law, rule). This term can be used in any branch of science *e.g.*, Taxonomy of plants, Taxonomy of Animals, Taxonomy of rocks etc. This means in this chaotic world of objects, taxonomy brings an orderly classification of all the things.

The term taxonomy was first introduced to plant science by A.P. de Candolle in 1813. According to him Plant Taxonomy means theory of plant classification.

The term taxonomy is based on the term taxon. The word 'taxon' was first used by a German Biologist Adolf Meyer in 1926 for animal groups. In Botany it was proposed by Herman J. Lam in 1948.

According to Radford *et al.* (1974) taxonomy includes identification, characterisation and classification.

Phases of Development in Plant Taxonomy

According of Davis (1963) there are four phases in Plant Taxonomy. They are

1. Exploratory or Pioneer Phase

The discovery, description, naming, identification and classification of plants.

2. Consolidation or Systematic Phase

The synthesis, mostly based on gross morphology, of field and herbarium knowledge in the preparation of floras, manuals, monographs, and form-based classification systems.

3. Experimental or Biosystematic Phase

The analysis of breeding systems, variation patterns, evolutionary potential and pertinent work in the chemical, numerical, cytological, anatomical, embryological and palynological aspects of systematics.

4. Encyclopaedic or Holotaxonomic Phase

The analysis and synthesis of all information and types of data in the development of one or more classification systems based on evolutionary or phylogenetic relationships.

Valentine and Love (1958) recognised the first three phases while Davis and Heywood (1963) added the fourth one.

The first two phases, which are mainly descriptive and based on gross morphological features, correspond to the "Alpha" taxonomy by Turrill (1938). The last two phases correspond to the "Omega" classification by Turrill (1938).

Aims and Objectives

The three main aims of plant taxonomy are classification, identification and nomenclature.

Classification

Man has to distinguish objects around him. For this purpose he has to classify them. Early man classified objects around him as animate and inanimate, useful and unwanted, plants and animals etc. In plants he classified them as those that are useful to him and not useful to him. Then he classified them based on habit; herbs, shrubs and trees. When he found the sexual parts, the stamens and the ovary he tried to classify them based on these sexual parts. Later when the number of plants increased he tried to classify them based on Natural relationships. After the Theory of Evolution by Darwin Botanists are classifying the plants according to the phylogenetic relationships.

Identification

There are millions of species around man and these include several thousands of plant species. Man devised various methods to identify them. Modern botanists identify the plants with the help of keys.

Nomenclature

For communicating about the plants around him man has given names to plants. Since there are different languages botanists have devised a method of naming plants known as Binomial nomenclature. These names are in Latin.

Identification, nomenclature and classification are the three important aspects of plant taxonomy. These are the main aspects of modern taxonomy also. Evidences obtained from different branches and their utilization is the important part of modern plant taxonomy.

Chapter 4
Concept of Population

Phenotype

Each individual organism consists of genotype and phenotype. The total outward (external) appearance is called phenotype. The total expressed characters or observable total form of an organism is called phenotype. Phenotype is the form or appearance of an individual and represents the result of external factors on its genotype. A phenotype is any observable characteristic or trait of an organism such as its morphology, development, biochemical or physiological properties or behaviour. Phenotype results from the expression of an organism's genes as well as the influence of environmental factors and possible interactions between the two.

Genotype + environment + random variation $\rightarrow$ phenotype.

Word origin: NL *phaeno-* < Gk *phaino-* shining, comb. form of *phaínein* to show, appear + -type.

The phenotype is the descriptor of the *phenome*, the manifest physical properties of the organism, its physiology, morphology and behaviour.

Genotype

This is the "internally coded, inheritable information" carried by all living organisms. This stored information is used as a "blueprint" or set of instructions for building and maintaining a living creature. These instructions are found within almost all cells (the "internal" part), they are written in a coded language (the genetic code), they are copied at the time of cell division or reproduction and are passed from one generation to the next ("inheritable"). These instructions are intimately involved with all aspects of the life of a cell or an organism. They control everything from the formation of protein macromolecules, to the regulation of metabolism and synthesis.

Stebbins (1950) explained genotype as sum total of all the genes present in the individual. The genotype of an organism is the inherited instructions it carries within its genetic code. Not all organisms with the same genotype look or act the same way, because appearance and behaviour are modified by environmental and developmental conditions. Similarly, not all organisms that look alike necessarily have the same genotype.

The "internally coded, inheritable information", or *Genotype*, carried by all living organisms, holds the critical instructions that are used and interpreted by the cellular machinary of the cells to produce the "outward, physical manifestation", or *Phenotype* of the organism.

Thus, all the physical parts, the molecules, macromolecules, cells and other structures, are built and maintained by cells following the instructions given by the genotype. As these physical structures begin to act and interact with one another they can produce larger and more complex phenomena such as metabolism, energy utilization, tissues, organs, reflexes and behaviours; anything that is part of the observable structure, function or behaviour of a living organism

The genotype is the descriptor of the *genome* which is the set of physical DNA molecules inherited from the organism's parents

The concepts of phenotype and genotype also demand the distinction between *types* and *tokens*. As the words "genotype" and "phenotype" suggest, these are types, sets of which any given organism and its genome are members, sets defined by their physical description. Any individual organism and its genome are members of those sets, tokens of those types.

The distinction between genotype and phenotype was introduced by Wilhelm Johannsen in 1908 as a consequence of the realization that the hereditary and developmental pathways were causally separate. This claim had already been made explicitly by August Weismann at the end of the nineteenth century, who differentiated between the *germplasm* of an organism, the tissue that forms the gametes to produce the next generation, and the *somatoplasm*, the tissues of the rest of the body. According to Weismann the somatoplasm developed and was influenced by the environment, whereas the germplasm was segregated early in development and was not susceptible to environmental influences. Thus, there could be no inheritance of acquired characteristics. Johannsen's distinction between genotype and phenotype, was however, induced, not by Weismannism, but by the rediscovery in 1900 of Mendel's work on inheritance in the garden pea.

The critical feature of Mendel's result was the result he obtained in the first and second generations of crosses between pea plants with clear-cut phenotypic differences. When a pure breeding red-flowered variety was crossed to a pure breeding white-flowered form, all the offspring in the first generation were red-flowered. When, however, these red-flowered hybrids were crossed with each other, both red-flowered and white-flowered plants appeared in the progeny. In order to explain this extraordinary reappearance of white-flowered plants in the second generation despite the fact that the first generation cross produced only red-flowered plants, Mendel distinguished between the internal state of the plants and their outward appearance.

He postulated the presence of internal discrete elements, "factors", that were contributed by the parents to the offspring. While these factors interacted in some way to produce the external appearance of the plants, they did not physically blend or contaminate each other, but maintained their discrete individuality. Thus a pure bred white-flowered plant had two white factors, one contributed by its maternal parent and one by its paternal parent, while red-flowered plants had two red factors. The hybrid between these two pure bred varieties would then have one white factor and one red factor. The red flowers of this hybrid were a consequence of the "dominance" of red factors over white factors in their causal interaction in producing flower color. That dominance in physiological action, however, in no way affected the nature of the factors themselves, which separated again, uncontaminated, when the hybrid plants produced pollen and ovules. As a consequence, when two hybrid plants were crossed, some of the offspring would have received a white factor from both pollen and ovules and would have white flowers.

Biotype

Biotype consists of all the individuals having the same genotype (Stebbins, 1950). The biotype in cross fertilized organisms usually consists of a single individual. But in self fertilized plants, the individuals may become completely homozygous and produce by selfing a progeny of individuals all within the same genotype and so belong to biotype.

Plasticity of Phenotypes

Phenotypic plasticity is the ability of an organism to change its phenotype in response to changes in the environment. Such plasticity in some cases expresses as several highly morphologically distinct results; in other cases, a continuous norm of reaction describes the functional interrelationship of a range of environments to a range of phenotypes. The term was originally conceived in the context of development, but is now more broadly applied to include changes that occur during the adult life of an organism, such as behaviour.

Organisms may differ in the degree of phenotypic plasticity they display when exposed to the same environmental change. Hence, phenotypic plasticity can evolve and be adaptive if fitness is increased by changing phenotype. In general, sustained directional selection is predicted to increase plasticity in that same direction.

Some responses will be similar in all organisms, for example in organisms that do not thermoregulate, as temperatures change lipids in the cell membrane must be altered by creating more double bonds (when temperatures decrease) or removing them (when temperatures increase).

Generally phenotypic plasticity is more important for immobile organisms (*e.g.*, plants) than mobile organisms (*e.g.* animals). This is because immobile organisms must adapt to their environment or they will die, whereas mobile organisms are able to move away from a detrimental environment. Examples of phenotypic plasticity in plants include the allocation of more resources to the roots in soils that contain low concentrations of nutrients and the alteration of leaf size and thickness. The transport proteins present in roots are also changed depending on the concentration of the nutrient and the salinity of the soil. Some plants, *Mesembryanthemum crystallinum* for example, are able to alter their photosynthetic pathways to use less water when they become water- or salt-stressed.

Nevertheless, some mobile organisms also have significant phenotypic plasticity, for example *Acyrthosiphon pisum* of the [Aphid] family exhibits the ability to interchange between asexual and sexual reproduction, as well as growing wings between generations when plants become too populated.

The familiar formula Genotype + Environment → Phenotype implies that the phenotype is not merely manifestation of the genotype, but that environmental factors play a part in modifying the latter to produce the former. Turesson referred to different phenotypes which were merely the product of differening environments as ecophenes, and Clements called them ecads. The ability to express a genotype as different phenotypes according to external conditions is referred to as Phenotypic plasticity or one may refer to plastic responses. It is important for taxonomists to recognize plasticity and to decide whether or not the different ecophenes should be taxonomically named.

Work on plasticity formed some of the earliest biosystematic experiments. In 1902 J Massart showed that the different growth-forms of *Polygonum amphibium* were adaptations depending upon whether the plant was growing in water or on dry land (or in intermediate conditions), as they could be readily interconverted by transplanting. G. Bonnier carried out a wide range of cultivations, starting in 1884, in Paris, the Alps and the Pyrenees, and claimed to demonstrate that a number of species could be interconverted according to the altitude at which they were grown, for example *Lotus alpinus* and *L. corniculatus*. However, some of Bonnier's conclusions are now known to be ill-based, the apparent convergence in characters of separate species often involving superficial features only, or being based on faulty experimental technique. Nevertheless, more sophisticated experiments along similar lines were later carried out by A. Kerner and by F.E.Clements, the latter leading to the long-term studies of Clausen, Keck and Hiesey.

In India *Tridax procumbens* population grows very well luxuriously in the rainy season with several leaves reaching the large size of 26 to 36 cm or so. In receding monsoon the surviving plants produce small sized leaves and stems to minimize the transpo-evoporation. Thus the plant population which have been surviving in winter parsimoniously spend water and tide over the adverse climatic conditions such that the large sized leaves in rainy season and very small sized leaves in the winter season. The possession of different types of different sized leaves are adaptations to the existing environment in favourable and adverse climatic conditions respectively. The response of different features of the plant to different environmental conditions or factors varies considerably.

Factors Affecting Phenotypic Variations and their Significance

Certain environments are well known to give rise to extreme ecophenes in a wide range of species, for example adaptations to shady/sunny, alpine/lowland and wet/dry conditions. Certain genera or species are also notorious for producing a wide range of ecophenes, and the taxonomist has to be on the look-out for them. In

Epilobium sun-plants have small, thick leaves, much anthocyanin, many hairs, and a short stature, whereas shade-plants have the opposite characters. Since one of the most important diagnostic characters in *Epilobium* is the type of indumentum, it is vital to examine the quality of the hairs, not their quantity, when making determinations. In many annual grasses and other plants dry conditions promote a dwarf habit, a normally tall plant with a well branched inflorescence often appearing as a midget a few centimeters high with a single spikelet or flower. Fortunately the measurement of the individual parts of the latter are usually not modified, though they may be.

Plasticity is extremely common in bryophytes, for example the genera *Hypnum*, *Sphagnum* and *Scapania*, where a great number of taxa have been described based on them. In the 1920s, H. Buch pioneered the study of plasticity in bryophytes, conducting cultivation experiments on a range of liverwort genera. Many examples are known also in marine, fresh-water and terrestrial algae and in lichens; in the latter case the nature of the substrate (*e.g.*, rock or tree-trunk) can be the important factor.

Plastic responses are by no means confined to exomorphology, for chemical and anatomical characters are frequently affected. The great variation shown by some species of *Eucalyptus* in terms of essential oil content, leading to the recognition of morphologically identical chemical races, is well-known as is the presence of cyanogenic and non-cyanogenic races of *Trifolium repens* and *Lotus corniculatus*. In the genus *Xanthium* (Asteraceae) the different ecotypes of one species have been shown to vary in their sesquiterpene lactones. Cryptic chemical races or *Concocephalum conicum* with different flavonoids occur in Europe and America, and many more examples of this kind of variation are coming to light.

In other instances the variation is not disjunct but continuous. For example, Flake *et al.* (1969) analysed samples of *Juniperus virginiana* every 150 miles along a 1500 mile north-east/south-west transect from Washington D.C. to Texas, U.S.A., for the presence of a wide range of terpenoid compounds, and detected a gradual change in terpenoid composition along the transect. Adams and Turner (1970) made similar studies on the related *J. ashei* throughout its range of distribution in Texas, and found that the more peripheral populations showed greater divergence from the population centre. These studies are particularly interesting because *J. ashei* and *J. virginiana* had been the subject of much detailed morphological and statistical work (Hall, 1952), which had strongly indicated that the cause of at least some of the variation in *J. virginiana* was its hybridization with *J. ashei* at and towards their point of contact in Texas. The clinal variation of *J. virginana* (which is also apparent in morphological features) was attributed to the greater and greater dilution of the effects of hybridization (*i.e.*, the lower and lower frequency of *J. ashei* genes) away from the south-western limits of *J. virginana*. This dilution effect, which has many well substantiated examples, is known as introgression.

Certain genera of flowering plants adopt different phenotypes according to the time of year at which they germinate and flower – the so-called seasonal variants. This phenomenon, often referred to as seasonal polymorphism, is particularly notable in montane taxa, which may be subjected to different lengths of season and temperature regimes according to the altitude at which they grow and the agricultural practices to which they are subjected; *Gentianella, Melampyrum* and *Rhinanthus* are good examples. Since these variants are often separated spatially they have mostly been given different names, frequently at the species level. In modern floras they are often treated as varieties or subspecies, but in most cases they have not been investigated properly (the three above are difficult to cultivate) and are often probably just ecophenes.

It has been pointed out by several workers that plasticity and ecotypification are alternative adaptive strategies which are both important in evolution.

Particularly perplexing are cases where certain ecophenes can mimic genuine ecotypes, for example dwarf variants of *Prunella vulgaris* adapted for growth in close-cut lawns (Nelson, 1965) or of *Cytisus scoparius* adapted for growth on exposed maritime shingle (Gill and Walker, 1971). In such examples the variant might be a genetically determined dwarf or a dwarf ecophene, and only experimental cultivation will distinguish the two.

The majority of taxonomists are of the opinion that ecophenes should not be given taxonomic status. There are a great many taxa (mostly infraspecific) based upon ecophenes, and when a taxonomist discovers their background they are usually no longer recognized as distinct, but are relegated as mere synonyms. Inevitably this process of discovery is slow, as most species have not yet been systematically cultivated. Probably the rate of description of new taxa which are in fact ecophenes still exceeds the rate of their relegation to synonymy. This is certainly so where little biosystematic research has been undertaken, particularly in the tropics and in lower plants.

Although in vascular plants there is almost universal acceptance that ecophenes should not be taxonomically named, there are some exceptions; Jones and Newton (1970) treated *Puccinella pseudodistans* as *P. fasciculata* form a *pseudodistans* because their experiments indicated that it is a growth form of *P. fasciculata* induced by waterlogged soils (although others dispute their results). Bryologists realize that the proportion of infraspecific taxa (and even species) which are ecophenes is far higher than in vascular plants, but nowadays there is some agreement that where detected, they should lose their status (Smith, 1978). In the algae the situation is more difficult still, because of the lack of experimental evidence. The lichens are of interest since lichenologists have for many years deliberately given names to ecophenes, usually at the level of the forma, although even in this group such a practice is being abandoned (Poelt, 1973).

Ecophenes are entities of a quite different nature from genetically determined variants, but they can be conspicuous in nature and they are of ecological importance. On many occasions it can be advantageous to refer to them concisely, *i.e.*, by means of a name, and the system developed by Buch (1922-28) appears to be very suitable for this purpose. In this system (Table 4.1) each ecophene is referred to as modification (abbreviation **mod.**) and those characteristic of each habitat receive the same name. There is, therefore, no question of any recourse to a code of nomenclature, or to adopting the earlier available name, as the selection of names is automatic. Buch's system was developed for liverworts; for other groups it would need extension, or perhaps replacement by a more appropriate series of names, and the ambiguous term, modification' would also be better replaced by a more precise one.

Table 4.1: The Categories Proposed by Buch (1922-28) for the Recognition of Ecophenetic Variants of Liverworts

	Modification	*Description and Cause*
1.	Parvifolia	Small leaves: in faint light
2a.	laxifolia	Distant leaves: in faint light and moist air
2b.	densifolia	Dense leaves: in dry air
3a.	leptodema	Thin cell-walls: in very moist air or water
3b.	pachyderma	Thick cell-walls: in dry air
4a.	viridis	Green, cell-walls colourless: in diffuse light
4b.	colorata	Red, brown or purple cell-walls, obscuring the chlorophyll: in direct sun-light
	Combinations, *e.g.* mod. Pachyderma-viridis, may also be used	

Chapter 5
Concept of Character

Definition of a Character

Overall similarities can be judged by a combination of characters. The natural selection operate on these combinations of characters. But it is convenient to deal with unit characters which can be treated quantitatively. The choice of such unit characters is largely a matter of experience. This naturally raises the problem of definition of unit characters. However, Davis and Heywood (1963) have attempted to give a general definition, which may be considered quite satisfactory in the absence of any suitable one. According to the author, a character is "any attribute (or descriptive phase) referring to form, structure, physiology or behaviour which is considered separately from the whole organism for a particular purpose, such as comparison, identification or interpretation'. In more general terms 'any expressed attribute' of the organism that can be measured, counted or otherwise assessed may be called a character.

The terms character, character-state and characteristic have often been used loosely and interchangeably. Strictly speaking, they denote different things. Character is the particular attribute that the taxonomists consider for classificatory purposes and is a mere abstraction, whereas the various patterns of their expressions are called character-states, Thus, the same character may have several character-state (Table 5.1).

When a particular character-state is exclusive to a particular taxon in a given assemblage, it is said to be 'characteristic' of the group. Thus bladders and pitchers are 'characteristic' of the insectivorous taxa *Utricularia* and *Nepenthes* respectively, while the leguminous fruit is 'characteristic' of the family 'Leguminosae'. Within the Leguminosae, the papilionaceous corolla is characteristic of the Papilionoideae; among the Lamiales, a gynobasic style is characteristic of the Labiatae; and so on.

Table 5.1

Sl.No.	*Character*	*Character-state*
1.	Leaf arrangement	Alternate, opposite, whorled
2.	Floral symmetry	Regular, zygomorphic, asymmetric
3.	Ovary position	Hypogynous, epigynous, perigynous
4.	Stylar position	Terminal, gynobasic
5.	Placentation	Axile, parietal, free-central, basal etc.
6.	Pollen aperture type	Sulcate, colpate, colporate, porate, etc.

Analytic vs Synthetic

In taxonomic procedure characters may be employed for two main activities: (i) identification, characterization and delimitation of species, and (ii) classification of these species into higher taxa. Corresponding with these activities, characters may be regarded as analytic or synthetic. There are diagnostic or key characters used for identification and characterization. These are often restricted in their occurrence so that they alone are enough to reach a correct diagnosis. They are most useful and easy for the process of identification and must be included in the floras and manuals.

Analytic characters which are mostly diagnostic ones, are useful in identification, characterization and delimitation of lower taxa, as stated above. On the other hand, synthetic characters are useful in grouping these taxa into higher groups. Synthetic characters gradually decrease in number as one goes from lower to higher groups. This is obvious because only a few characters become increasingly constant at higher position of the group in the hierarchy. The characters such as sympetalous versus polypetalous condition, superior versus inferior gynoecium, etc. are quite constant and do not vary within smaller groups or within populations of the same species. They are also termed as constitutive or organizational characters.

Table 5.2

Sl.No.	*Character*	*Analytic*	*Synthetic*
1.	Enclosed ovules	Spermatophytes	Angiosperms
2.	Parallel venation	Angiosperms	Monocotyledons
3.	Latex	Compositae	Asclepiadaceae
4.	Capitate inflorescence	Rubiaceae	Compositae
5.	Leguminous fruit	Polypetalae	Leguminosae

Qualitative vs Quantitative Characters

Characters that can be measured or counted are termed quantitative characters, *e.g.*, leaf size, number of stamens, seed number, relative length of stamens. Those that can not be measured are qualitative characters, *e.g.*, flower colour, leaf arrangement. In taxonomy both types of characters are used. Qualitative characters are generally more useful to distinguish taxa of specific or higher ranks while quantitative ones are often useful to separate lower taxonomic categories at infra-specific levels. In taximetrics, however, quantitative characters are preferred, though attempts to apply different values to qualitative character-states in the computation of similarities in numerical techniques are also being made.

Homology vs Analogy

Phyletic (=phylogenetic or evolutionary) characters are used primarily in phylogenetic (=phyletic) classification. The most important distinction between characters that are homologous versus those that are analogous. The difference between these terms is on the surface simple, but many problems exist philosophically. Richard Owen's original definitions were, for homologue, "the same organ in different animals under every variety of form and function, "and, for analogue "a part or organ in one animal which has the same function as another part or organ in a different animal." Homologous characters, therefore, were originally viewed prior to Darwin's theory of evolution as simply basic structural differences. After evolutionary theory developed, homologues were viewed as the structural modifications of the same organ, inherited from a common ancestor, in response to different selection processes. Analogues, on the other hand, were those features developed by different organs to the same selection processes. There is no question that for the proper construction of evolutionary classification, homologous characters need to be emphasized. The detection of homologous characters in two groups is done by knowing that they have descended from a common ancestor. If this were known, however, there would be no need to use these homologues to reconstruct the phylogeny; it would be known beforehand in order to select the homologous characters. This circularity is a problem which has led some workers to eschew searches for homologous characters (Davis and Heywood, 1963). Philosophically there are solutions to the problem (Ghiselin, 1966; Hull, 1967; Cain, 1976), such as by considering other features of the organisms for additional signs of similarity (Cain, 1976), but they boil down simply to the reliance on structural and ontogenetic similarity as a reflection of homology. The detection of homologous characters is a difficult problem for any phylogenetic reconstruction and pitfalls can occur (see the problem of interpretation of leaf homologies in *Acacia*, in Kaplan). Because of the complexities of the issue, botanists have tended to deal with the problem obliquely, as indicated in the recent text by Stace: "Homology is usually defined on the basis of common evolutionary origin, a definition which should in theory be uncontentious, but which in fact is usually quite impractical because of our lack of evolutionary data. In practice, therefore, one can only guess at homologies by making as detailed as possible and investigation of the structures concerned. More usually the problem is ignored".

Consistent vs Variable

Although theoretically all characters are useful in classification, their use, as often experienced, appears to be limited to a small number. This is because certain characters are considered to be good or more reliable than others. Good characters are those that: (i) are not subject to wide variation with in the samples being considered; (ii) do not have a high intrinsic genetic variability; (iii) are not easily susceptible to environmental modification and show consistency, *i.e.*, agree with the correlations of characters existing in a natural system of classification. All these conditions may be summed up by stating, "a taxonomic character is only as good as its constancy".

Heterobathmy

The problem of weighing would not be so difficult if all the characters of an organism evolved harmoniously, at an equal rate, and occupied the same level of the evolutionary development. But it is a well-known fact that different organs have evolved at different rates. This phenomenon of unequal rate of evolution of different features within one lineage is known as "mosaic evolution" (De Beer, 1954). Mosaic evolution results into different evolutionary stages or grades. The difference in these grades is termed heterobathmy (from greek bathmos = step, grade).

An organism may present a mosaic combination of characters of quite different evolutionary levels because of heterobathmy. Thus for example, the genera *Trochodendron, Tetracentron* and *Sarcandra,* with their primitive, vesselless wood have rather specialized flowers, where as the genus *Magnolia* which possesses a comparatively much more primitive type of flower, the wood is rather advanced with vessels having simple perforation plates.

The more strongly heterobathmy is expressed, the more contradictory is the taxonomic information provided by different sets of characters. In such cases, only the application of various methods can reveal those "critical characters" and "critical tendencies" (Wernham, 1912) which are reliable phyletic markers. Correct weighing of the characters and their evolutionary tendencies gain special significance in such cases.

Character Weighing

In organisms there are innumerable characters, and all of them cannot obviously be considered in taxonomy. However, when we start classifying, we do not take characters at random, instead, we start by selecting characters, usually based on two primary considerations

1. They should be useful diagnostic characters which do not vary within the group,
2. Such invariant characters are of no use in intragroup studies. For this purpose, characters that are variable within the group are selected.

When we select characters thus, in effect we are giving a greater importance (weight) to those selected. Heywood (1967) called selection weighting. In turn, many other characters are rejected by rejection weighting or residual weighting.

Weighting of characters is a technique in evolutionary taxonomy but it is not as new as the latter. Traditional taxonomists also weighted in an intutive manner and without even being aware of it, in constructing classifications (Baum, 1976). The Aristotelian concept of 'essence' gives overriding importance to characters that constitute the 'type' and rejects all those which are variable and not universal. John Ray who was followed recently by Hutchinson, considered that growth habit was of great importance in Dicotyledon evolution. Linnaeus classified the whole plant kingdom on sexual characteristics. In fact, most of the existing systems of Angiosperms are based primarily on floral characters rather than on vegetative ones.

Selection or rejection of characters is usually an intuitive process. Taxonomists select characters that are supposed to be of greater taxonomic importance, those which are genetically more stable and conservative, those that are of greater diagnostic value and those thought to be evolutionary markers. This type of selection of characters for their supposed significance 'before one has any evidence to justify it' is called '*a priori*' weighting of characters (Heywood, 1967). Van der Pijl and Dodson (1966) suggested that vegetative attributes are somewhat less important than reproductive characters in taxonomy. This *a priori* thinking may be true with regard to certain superficial attributes such as leaf size and shape and pubescence, which are very highly plastic and rarely of use at higher levels of classification. But such criticism applies to certain reproductive characters, too. Clifford and Laverack (1974) have found that, in the case of Orchidaceae, either of these types of character (reproductive or vegetative) can 'mirror the phylogenetic history of the family more faithfully than the other.'

However 'a priori' weighting of characters still continues to hold sway in the various sub disciplines of taxonomy, each of which gives overriding importance to the characters derived from its respective field of investigation while giving less attention to or even ignoring those from other areas. Consequently we have different taxonomies, *e.g.*, cytotaxonomy, chemotaxonomy, instead of a 'synthetic' general-purpose classificatory system based on overall similarities and dissimilarities.

In evolutionary taxonomy, botanists believe that all characters are not of equal importance in determining the phylogenetic history of a given group. However, there can be considerable disagreement as to the relative importance of characters (Stebbins, 1974). The concept of 'ground plan' comparable with the Aristotelian 'essence', is largely based on such considerations and hence is, to a great extent, aprioristic.

Mayr (1969) defined weighing as 'a method for determining the phyletic information content of a character'. He also suggested that complex characters, shared derived features (synapomorphies), constant and consistent characters that are not affected by ecological shifts and those displaying high correlation might be given a high rating. Polygenic characters, having higher information content with regard to other characters, are superior to monogenic or oligogenic ones, which are of low value). Thus there is evidently much room for '*a priori*' considerations.

Michel Adanson (1727-1806) and his empirical school rejected this system of providing different weighting for different characters as a prelude to classification and replaced it with an empirical process called *a posteriori* weighting. Hence, all characters are considered to be of equal weight to begin with and classification-conducted is based on a similarity-dissimilarity ration (similarity co-efficient). The relative importance of each character is revealed only by detecting and estimating the degree of correlation with other characters, the rating varying directly with the degree of correlation. This type of weighting is also called as *correlation weighting*.

A *posteriori* weighting of characters is today strictly employed only in numerical taxonomy. This envisages a large body of equally weighted characters (descriptors) and leads to deletion, restructuring and *a posteriori* weighting of these descriptors through recognition of the main character cluster in a diagram (Legendre, 1975; Hogeweg, 1976). The differential rating estimated *a posteriori* is then used in circumscribing and delimiting the taxa involved. With the introduction of numerical methods in evolutionary taxonomy, Farris (1967) has suggested character weighting to detect conservatism and convergence. In evolutionary systematic, different characters with different correlation values are allocated differential phyletic weighting depending on the phyletic information-content. But estimating the latter is an extremely difficult task, especially when evolutionary reversals and convergences occur.

Despite what has been said about the need for equal rating for all characters, it remains extremely difficult to devise fool-proof methods to estimate *a priori* subjective considerations. In taxometrics, computational analysis serves this purpose to a large extent, but the initial selection of characters, their definition and formulation, and the demarcation of taxonomic groups are all based on 'individual subjective judgments of the taxonomist (Baum, 1976), and naturally this will be reflected in the resulting classificatory system. It seems that it is impossible to escape a moderate degree of subjectivity in human affairs, including classification.

The problem of weighting of characters has recently been the subject of intense discussion, and has proved to be a specific taxonomic problem that can be solved only by the systematist (Takhtajan, 1980). To add to his difficulties in this endeavor there are evolutionary phenomena such as heterobathmy and the mosaic evolution that results from it. In one group of a given taxon, certain characters will have evolved to a greater degree than other, whereas in the related group the reverse will be the case. This phenomena of differential evolutionary rates, which was termed heterobathmy by Takhtajan, results in the appearance of a mosaic of variously primitive and advanced stages of a character over the whole taxon. In highly heterobathmic groups, study of individual characters produces conflicting data so that a evaluation of such groups becomes another problem for the systematist. For example, while there is almost general agreement (based on paleobotanical data) that flowering plants had their origin during the Cretaceous, Ramshaw *et al.* (1972) suggested on the basis of protein studies that they might have originated as far back as the Ordovician. Where there are strongly divergent data, confusing the systematists about the evolutionary trends, more complete all-round studies are required to detect the 'critical characters' and critical tendencies' which are reliable phyletic markers. Unless and until these are discovered, weighting of characters will remain a problem.

Correlation of Characters

Characters and character-states are not distributed at random among plant and animal groups. Instead, they display various kinds of degrees of association, *i.e.*, certain characters show a tendency to cling together and are hence transmitted together. Such characters are called correlated characters. The concept of correlation of characters has a great bearing on our knowledge of the process of evolution and the systems of classification that we have now.

In the absence of enough fossil evidence, our knowledge of Angiosperm phylogeny is mainly dependent upon information from living taxa, which is usually then extrapolated into the past. One of the most important and commonest methods of recognizing ancestral and derived features is by 'taxonomic series analysis', where the evolutionary trend of a character in question is read in the direction of another which is already known or inferred, *i.e.*, by correlation studies. Furthermore, Sporne (1948) suggested that primitive characters tend to occur together. The corollary of this is that characters that are positively correlated with characters known to be primitive are also likely to be primitive. Thus, Bailey and his associates investigating the phylogeny of angiosperms found that the trilacunar node is common in groups that were thought by some botanists to approximate most closely to the earliest Angiosperms. They concluded that the trilacunar node is primitive. The later discovery of a correlation between the trilacunar node and stipulate leaves led to the conclusion that the latter are more primitive than exstipulate ones. Subsequent studies have revealed several such correlations. Sporne (1976, 1977) investigated the correlation of 26 characters among dicotyledons and on this basis calculated advancement indices of families before placing them in his classificatory scheme. He presented his classification of dicotyledons (1956) and monocotyledons (1976) in hypothetical circular schemes, the orders with the lowest advancement indices placed nearer the centre than those with higher indices towards the periphery. In short, the whole of the phylogenetic superstructure that we have now depends very much upon evidence of character correlations, and as such this concept is highly useful in evolutionary and systematic studies.

Recently, several workers have criticized this approach, suggesting that the correlation of characters is no more than a reflection of functional interdependence between them, and need not necessarily indicate evolutionary history, and that such adaptive characters, because they are liable to differ between closely related forms, are not useful for higher level classification. The adaptive value of one or more character combinations might be so cryptic or subtle that limited knowledge might not allow one to discover it. In fact, any character that persists for a long time in a phyletic line should be unaffected by natural selection and hence tends to be adaptive at a deeper level. Casual variations in such characters are thus less likely to be advantageous within the broad mode of life. Moreover, non-adaptive characters (such as leaves in completely parasitic plants and wings in flightless insects) are also likely to be extremely unstable and may vary greatly in related forms (Crowson, 1970). It seems that therefore that the functional interdependence and adaptiveness of characters do not have any bearing on their taxonomic usefulness. Sporne found that correlation exists also between such characters as the presence of stipules and nuclear endosperm among which any functional interdependence is very unlikely. 'The correlation between two such characters, separated as these are both in time and space during the growth of the plant, is much more likely to arise because both characters are primitive (Sporne, 1976).

Stebbins (1951) suggested that correlated characters cannot be interpreted as primitive or advanced; instead he suggested that one should look at the various 'syndromes' that occur more frequently than others in a given group and pointed out that some combinations are more efficient than others. But all such combinations are not necessarily advanced. It is likely that combinations of primitive characters would also make efficient combinations, just as the 'magnolioid' characters make efficient kinds of dicotyledon in the tropical rain forests (Sporne, 1977).

Modern taxonomy leans very heavily on multiple correlations rather than one single character differences. Defining monophyletic groups is especially helpful in the clear demarcation of taxa, for it is less likely that all such characters in a given taxon have arisen independently. Thus the number of cotyledons in angiosperms is correlated with the type of leaf, the orientation of vascular bundles, the nature of flowers, etc. In dicotyledons, the two cotyledons are generally correlated with dorsiventral, reticulately veined leaves, four- to five merous flowers and collateral open vascular bundles arranged in the form of a broken ring, whereas the single cotyledon in monocotyledons is usually associated with isobilateral parallel-veined leaves, three-merous flowers and closed and scattered vascular bundles. In the sympetalae the gamopetalous corolla is usually correlated with epipetalous stamens.

Correlation, however, is relevant only to the group under study. In a group X, the character 'a' is correlated with 'b', but it may be correlated with another character 'c' in taxon Y. The gynostegium is correlated with all moncotyledonous characters, zygomorphy of the flowers and an inferior ovary in the Orchidaceae, whereas the same coincidence with dicotyledonous characters along with actinomorphic flowers and a superior ovary in the Asclepiadaceae. The bilabiate corolla in the Labiatae is correlated with a gynobasic style. In Verbenaceae the very same type of corolla goes with the terminal style.

Chapter 6

Methods of Sampling and Processing Data

From a taxonomist's point of view, it is depressing to reflect upon the mass of published data that exists from plants dealing with cytology, chemistry, anatomy, and so forth, but which is of marginal use for gaining systemic insights because of the small sample size or lack of adequate documentation provided. In taxonomic studies, therefore, it is imperative to pay close attention to several considerations when gathering comparative data (Table 6.1).

Collection of Data

Sampling

The first consideration in gathering comparative data is sampling. This aspect of data-gathering is usually done in the field, and for this reason, the investigator must have a clear idea of what he or she is attempting to accomplish in the project before beginning field work. Three basic questions must be answerable before effective sampling can be done: (1) What parts of the plant should be sampled? (nature of sample); (2) What size should the sample be? and (3) What techniques should be used to collect the desired plant parts? The ability to answer satisfactorily these three questions relates directly to the ability of the worker to know where he or she is going geographically, how long the trip needs to be, what will be collected once there, and what kind of equipment will be needed for making the desired types of collections. To understand these ideas, the investigator must know clearly what type of systematic problem is being addressd, and what kinds of data will be needed to solve the problem. Some type of studies, such as those oriented primarily toward classification in a single genus, will require probably fewer population samples and fewer individuals per population than if intensive studies are being conducted on a closely-knit complex of taxa in which hybridization is suspected.

Table 6.1: Outline of Considerations when Gathering Comparative Data

I. Sampling (Field)
   A. Nature of Sample
      1. Parts of the plant to sample
      2. Additional features of the plant worth noting that will not be sampled
   B. Size of Sample
      1. Number of parts per plant to sample
      2. Number of plants to sample
      3. Number of populations to sample
   C. Techniques in Sampling: methods for collecting desired plant parts

II. Measurement (Laboratory)
   A. Nature of Measurement: parts of plant to measure
   B. Size of Sample to Measure
      1. Number of parts per plant to measure
      2. Number of plants to measure
      3. Number of populations to measure
   C. Techniques in Measurement methods for obtaining data

Note: Keep in mind that four factors will realistically control the nature and extent of any data-gathering attempt: (1) purpose of study; (2) time; (3) space and (4) money.

In general, a sample is adequate if it documents well the variability in character states at the next lower level in the hierarchy (whether formally recognized or not). For example, if relationships among genera are being investigated, the character variation among all the species within each genus should be known. If not all the species can be studied, then this does not mean that no comparisons among the genera can be made, but rather that the final interpretation of relationships will be less convincing. If closely related species are being examined, the intraspecific variation (at the subspecific, varietal, or population levels) in each should be documented well before comparisons are made. In problems at any level, as new data are collected, the amount of additional variation that is added by the new samples needs to be examined. If variation of large deviations from the inferred (or calculated) mean continues to be obtained, more data probably should be collected for a better understanding of the distribution of character-states before the final relationships are assessed.

Measurement

The second consideration in gathering comparative data is measurement. Although this aspect usually is completed in the laboratory, which gives the investigator more freedom from time restrictions, measurement is dependent up on plant parts collected in the field, and therefore, if the field sampling is inadequate for some reason, the measurements also will be incomplete. Before the taxonomist heads for the field, he or she must not only be able to answer these same questions relating to sampling but must also ask and be able to answer these same questions relating to measurement: (1) what parts of the plant to measure (nature of measurement); (2) how many parts should be measured and from how many individuals in how many populations (size of sample to be measured); and (3) what techniques will be used to obtain the desired data?

It is most important that the measurements be obtained in a consistent and acceptable manner. If the measurements are not properly made, the classifications based upon these data will be inaccurate or misleading. For measurements to be obtained consistently, the exact same type of equipment must be used in all cases to measure the exact same structures from precisely the same parts of the plants. For example, data on pubescence (hairs per square area) must be gathered using the same rule under the same magnification and from the same organs of each plant. For measurements to be gathered in an acceptable fashion, the worker must be thoroughly familiar with the type of data he or she is using. It must be certain that the techniques for obtaining the data will not cause alterations of the generated information. For example, some chemical data, such as essential oils (monoterpenoids), require delicate handling. If the compounds are not treated with great care (sealed vials, low temperatures, etc.) structural rearrangements can take place which when the data are recorded, might lead to a distorted view of the relationships.

In the final analysis, four factors realistically control the nature and extent of any data-gathering attempt: purpose of the study, time, space, and money. As is true for the majority of our activities whether scientific or otherwise, inevitably the ultimate limiting factor is money. Time and space relate directly to money, and even the purpose of a particular study is usually delimited indirectly by financial considerations. However, these realistic considerations do not necessitate that a taxonomist must keep problems so narrow that no difficulty is ever encountered in obtaining the needed data. Rather, one determines first what the problem of interest and significance is, then what the realistic requirements are, and finally what time, space and money are needed to gather the data to solve the problem.

Evaluation of Data

In the process of collecting data, and even after data have been collected, it is important to evaluate them for their potential for helping to solve the taxonomic problems at hand. Sometimes this evaluation can be done by simple inspection. Often, however, more sophisticated methods for evaluating these data are used, and these can be called techniques of mathematical analysis and summarization (Crovello, 1970). These can help find patterns and structure in the data even though such patterns may be difficult to see by simple visual examination. Many statistical approaches exist such as correlations, regressions, analysis of variance and covariance, and basic statistical measures of means, ranges and so on. Various similarity coefficients will be used in phenetics, cladistics, or with explicit phyletics, and many different algorithms can be utilized before the final approaches are selected for presentation in the published report. Information theory can also be used to help determine the robust quality of classification (*e.g.*, Duncan and Estabrook, 1976). The complex patterns of multidimensional variation can be reduced into fewer dimensions by ordination of different types, multidimensional scaling, and cluster analysis. Discriminant function analysis is also helpful, especially in situations involving hybridization, and other types of geometric and/or calculus evaluations also exist. In short, there is no lack of available sophisticated methods for data analysis to help make taxonomic decisions The paper by Crovello (1970) is most helpful, and the text of Sokal and Rohlf (1981) is also recommended as a starter in this area.

Clearly any particular set of comparative data should be examined rapidly at the earliest possible stage of a study for potential utility. If no variation exists in the groups with regard to a particular type of data, obviously this will be of no value whatsoever in helping to resolve relationships within that group. They may be extremely helpful, however, in delimiting that group from other groups at the next higher level in the hierarchy. The point is that sometimes it is difficult, if not impossible, to find taxonomically useful discontinuities within a particular set of data. This sometimes occurs even with the best of efforts by the taxonomist. The simple fact is that due to the dynamics of the evolutionary process, sometimes conditions are such that sharp discontinuities among taxa do not exist.

The evolutionary factors of speciation and divergence, which are responsible for the production of most of the diversity, at the same time may cause temporary (in an evolutionary sense) intergradations or continue to occur that obscure the usually observed discontinuities. A good knowledge of the evolutionary process, therefore, enables a taxonomist to approach more effectively and interpret more successfully, the existing patterns of relationships in these more challenging situations. It is certainly true that a person does not have to know anything about evolution to be a good taxonomist (consider all the excellent pre-Darwinian workers, such as Jussieu and Candolle), but it is also true that the acquisition of such additional knowledge will help make him or her a much better worker.

It is not within the scope of this book to elaborate in detail all aspects of the processes of plant evolution that may contribute to the obscuring of discontinuities. Rather, it is the objective to outline briefly these various evolutionary aspects and refer the reader to references that will clarify this understanding. Table 6.2 lists the major aspects of the processes of plant evolution with which the plant systematist should be familier to be well equipped to handle taxonomic problems. An excellent introduction to these processes is found in Stebbnins (1977).

Relative Efficacy of Different Kinds of Data

Many claims have been made regarding the power of different types of data for solving taxonomic problems. Whenever new information appears on the scene, a brief 'bandwagon' effect ensues, and the community passes through a vociferous period of advocacy only to settle down a decade or so later to a new type of integration of approaches. Such was the case with chemosystematics in the early 1960s, principally with secondary plant products and we now see it with regard to chloroplast and nuclear DNA. People sometimes point to the complex metabolic and development interactions that exist from DNA to the final expression of morphological

Table 6.2: Outline of the Major Aspects of the Processes of Plant Evolution (after Stebbins, 1977)

- I. Phenotypic plasticity
- II. Genotypic variation
  - A. Mutation (point mutation)
  - B. Recombination
  - C. Reproductive systems
    - 1. Asexual reproduction
      - a. Vegetative reproduction
      - b. Apomixis
    - 2. Sexual reproduction
  - D. Chromosomal aberrations
    - 1. Alteration of linkage groups
      - a. Translocation
        - i. Reciprocal
        - ii. Non-reciprocal
      - b. Deletions
      - c. Duplications
    - 2. Multiplication of genome
      - a. Autopolyploidy
      - b. Allopolyploidy
  - E. Ecotypic differention
  - F. Hybridization and introgression
    - 1. Hybridization
      - a. Diploid level
      - b. Polyploid level: allopolyploidy
    - 2. Introgression

traits, and suggest that the further back one goes toward the absolute genetic material, the DNA, the closer one comes to having the 'best' data for classification. From what is known of DNA at the present time, this is simply will not provide a panacea any more than will any other single source of data. One of the difficulties with nuclear DNA is that numerous sites are inactive and many feedback mechanisms exist. Hence, until we know more about the developmental interactions of the sequences, we will be unable to understand fully their evolutionary meaning. We certainly must push aggressively forward to obtain these data and understand their relevance, but we must also be ready to admit their limitations (as we also must always do with all other types of comparative data). A reasonable and balanced perspective on the positive value of both molecular and morphological data is given by Hills (1987).

Stuessy (1990) is of the opinion that in general sense no single type of data holds supremacy in determining relationships for purposes of classification. As A.J. Sharp

(1964) pointed out years ago: "Should any botanist think he has the final technique or the final answer, may I remind him that science has taught us nothing more clearly in this century that there are no absolutes and that everything is relative and can be predicted only within certain statistical limits". One might argue that if a person had to choose only one type of data for classification for use at all levels in the hierarchy, then probably morphology would be the wisest choice. The reason for this selection would be that in all probability the morphological attributes taken collectively would give the best indication of the evolutionarily significant features of the organism, and therefore, would be most useful for purpose of classification. However, the point to be stressed is that one rarely has to make such a decision. It is far better to remember that all types of data tell something about the genotype and adaptational and evolutionary history of the plants under study, and therefore, that all types of data should be used whenever possible. Obviously, every taxonomist has special training and interests, but whether by collaboration or by broadening one's perspective, an attempt should be made to bring as many different types of data to bear on a problem as possible. Only by this combined approach can the most useful and predictive classification of plants result. As Wagner has quipped, to deal with all these different types of data, the skilled taxonomist really needs to be a 'chemo-cytomorphotaxonometrical' to be effective.

Chapter 7
Breeding Systems

Many of the evolutionary properties of populations depend up on the heterozygosity maintained by a relatively steady rate of recombination within the population. However, in outbreeding populations chromosomal differences (*e.g.*, inversion, translocation), assertive mating, and, occasionally, varying amounts of intrapopulation sterility provide a certain degree of restriction to complete or random recombination. Many flowers are normally bisexual or hermaphroditic, thus inbreeding at the level of the individual is possible for most higher plants. Conversely, in these plants capable of self-pollination there are found a number of mechanisms that provide for some recombination through either frequent or occasional oucrossing. Since both inbreeding and outcrossing each have survival and evolutionary value under different conditions of natural selection, it is not surprising to find that many plants have evolved slightly ambivalent reproductive systems that allow for the effective operation of both of these divergent breeding mechanisms within a single population, or, in some cases, a single individual.

In general three different systems, based on the degree of inbreeding versus the degree of outbreeding, can be recognized in plants: (1) predominant outcrossing, (2) predominant selfing, and (3) mixed selfing and outcrossing. In addition, several highly successful methods of asexual reproduction have evolved among some of the higher plants and the combination of sexual and asexual reproduction adds to the evolutionary flexibility (and often the range of phenotypic variation and taxonomic confusion) of those plant groups in which it occurs.

In dioecious plants, where one plant bears only male or staminate flowers and another plant bears only female or pistillate flowers, we find maximum obligatory cross-fertilization. However, the enforced immobility of plants makes strict dioecism of dubious survival value, especially in a small population which might consists of plants of only one sex. Selection might therefore favour the production and survival of plants that are functionally diecious but which may either normally, or under certain environmental conditions produce a few perfect flowers. Such plants would be polygamodioecious. If some perfect flowers occur on an otherwise female plant the plant would be gynodioecious. The perfect flowers (as well as some of the pistillate flowers) would be the result of normal cross pollination from another, staminate plant. The gynodioecious condition, which occurs in some species of the pink, mint and aster families (Caryophyllaceae, Lamiaceae, Asteraceae) allows for a maximum of recombination and a minimum of homozygosity.

In populations of sexually reproducing organisms maximum genetic recombination, or gene flow, would occur if the population is panmictic, that is, if all individuals in the population could mate at random. This means, that any one individual is equally likely to mate with any other individual of the opposite sex in that particular population. Although panmixis is a useful evolutionary concept for the theoretical calculations, or as a reference point in experimental work, true panmixis most likely does not occur, or if so, only rarely, in natural populations of higher organisms. Among plant populations panmixis might be most closely approached in a population of wind pollinated plants. However, interplant distances, wind velocity, and the prevailing wind direction would all be factors that would reduce the randomness of wind pollination and therefore the randomness of interbreeding among the individuals comprising the population.

I. Outbreeding Systems

An out breeding system is one in which sexual reproduction involves the mating and union of gametes of different individuals.

Plants can be predominantly out breeding, or some mixture of the two. In many flowering plants specific mechanism has evolved that promote one of these systems. Outbreeding, also called out crossing, allogamy or xenogamy is the transfer of gametes from one individual to another genetically different individual.

In plants, out breeding may be obligate or facultative. The necessity or tendency for out breeding may be reinforced by dioecism, self incompatibility, heterostyly and other mechanism. Out breeding is usually by the wind, insects, or other pollinating agents,. Out breeding may imply mating between individuals that are less closely related than would occur in random mating, as in panmictic populations in which each individual has the probability of mating with any other individual.

The inability of functional male and female gametes to effect fertilization in particular combinations is known as Incompatibility. This is due to the arrest of post-pollination events at different levels. Incompatibility occurs between species (interspecific incompatibility) as well as within the species (intraspecific incompatibility). Whereas interspecific incompatibility prevents fertilization between gametes of distantly related species, intraspecific incompatibility prevents fertilization between gametes of the same or other individuals of the same species. Intraspecific incompatibility is also called as Self-incompatibility.

Interspecific incompatibility is heterogenic *i.e.*, controlled by more than one gene at different loci on the chromosomes (Linskens, 1975). It prevents free cross pollination.

In incompatible interspecific crosses either fertilization does not occur or syngamy is followed by abortion of hybrid embryo due to inadequate development of endosperm or embryo-endosperm incompatibility.

In nature different floral adaptations, such as dichogamy, herkogamy and unisexuality have evolved to prevent self pollination but the most widespread and effective natural device to enforce outbreeding is self-incompatibility.

Dichogamy

Maturation of male and female sex organs at different times is called dichogamy. It ensures cross pollination and prevents self-pollination. Dichogamy is of two types (a) Protandry: Here anthers dehisce much before the stigma of the same flower attains receptivity, *e.g.*, Maize, *Saxifraga, Helianthus, Phlox, Impatiens.* (b) Protogyny: If the stigma loses its receptivity, by the time the anthers dehisce, it is called protogyny, *e.g., Pennisetum, Aristolochia, Helleborus, Annona, Prosopis, Solanum.*

Herkogamy

In a bisexual flower, if both sex organs mature at the same time, the self-pollination is prevented by the arrangement of male and female reproductive organs at different heights (*Hibiscus*) or projected in different directions (*Gloriosa*).

Unisexuality

Unisexuality or dicliny is a condition in which the flowers are either staminate (male) or pistillate (female).

(a) Monoecy

Staminate and pistillate flowers occur in the same plant, either in the same inflorescence, *e.g.*, castor, mango (*Mangifera indica*), banana (*Musa sapientum*), and coconut or in separate inflorescences, *e.g.*, maize. Other monoecious species are cucurbits (*Cucurbita* sp.), Walnut (*Juglans regia*), chestnut, straw berries (*Fragaria* sp.), rubber (*Hevea braziliensis*), grapes (*Vitis vinifera*) and cassava (*Manihot utilissima*).

(b) Dioecy

The male and female flowers are present on different plants, the plant in such species are either male or female, *e.g.*, Papaya (*Carica papaya*), date, hemp, asparagus and spinach. In general sex is governed by a single gene, *e.g.*, asparagus and papaya. In some cases there are hermaphrodite flowers in addition to male and females, and a number of intermediate forms may also occur.

Self-Incompatibility

It refers to the failure of pollen from a flower to fertilize the same flower or other flowers on the same plant. More than 300 species belonging to 20 families of Angiosperms show self-incompatibality.

Intraspecific incompatibility is of two types

1. Heteromorphic (heterostylous), and
2. Homomorphic.

(a) Heteromorphic Incompatibility–Heterostyly

When plants of the same species produce more than one morphological types of flowers, it is called heteromorphic incompatibility. Different plants of each species may produce two (dimorphic/distylous) or three (trimorphic/tristylous) types of flowers, but each individual produces only one type. The relative levels of the stigma and stamens in different forms, and their incompatibility relationships are diagramatically represented in Figure 7.1. Distyly and tristyly together is referred to as Heterostyly.

Distyly

In this type two types of flowers are produced - long styled, also known as pin morphs and short-styled or thrum morphs (Figure 7.1A). Pollination is successful between individuals of different morphs (intermorph). Self-pollination and pollination between plants of the same morph (intramorph) are incompatible. Distyly is common in several taxa of the families, Rubiaceae, Plumbaginaceae, Linaceae and Boraginaceae (Bir Bahadur, 1978) and *Primula.*

Distyly is controlled by a single gene complex, S with two alleles S and s. The allele for short-style (S) is dominant over the allele for long-style (s). Short styled plants are heterozygous (Ss) and long styled are homozygous recessive (ss). Incompatibility in pollen is sporophytically determined. Pollen with 's' allele form short-styled form is compatible on long-styled pistil (ss), although incompatible on short-styled pistil (Ss).

Tristyly

In this type the plants produce three types of flowers - long-styled, mid-styled and short-styled (Figure 7.1B). Each morph produces anthers at two different levels, which correspond to the level of stigma of other two forms. For a successful pollination pollen grains are required not only from flowers of other morphs, but also from anthers corresponding to the same level as that of the stigma (Figure 7.1). For example, in short-styled morph pollen from long stamens are compatible on pistil of long-styled morph and not on mid-styled morph, those from mid-stamens are compatible on pistils of mid-styled morph and not on long-styled morph. Although most investigators limit the number of families with tristyly to four, Bir Bahadur (1978) considers tristyly to be present in 57 genera belonging to 14 families. Some of the examples are *Oxalis, Lythrum, Narcissus* and *Eichhornia.*

Tristyly is determined by two genes M and S, with two alleles each, S being epistatic to M. Long styled morphs are homozygous recessives for both genes (ss, mm), mid-styled morphs are homozygous recessive for s and homozygous dominant or heterozygous for M (ss MM/ss Mm), and short-styled morphs are heterozygous for S while M may be in any form (Ss mm/Ss Mm/Ss MM).

In both distylous and tristylous plants pollen incompatibility is sporophytically determined.

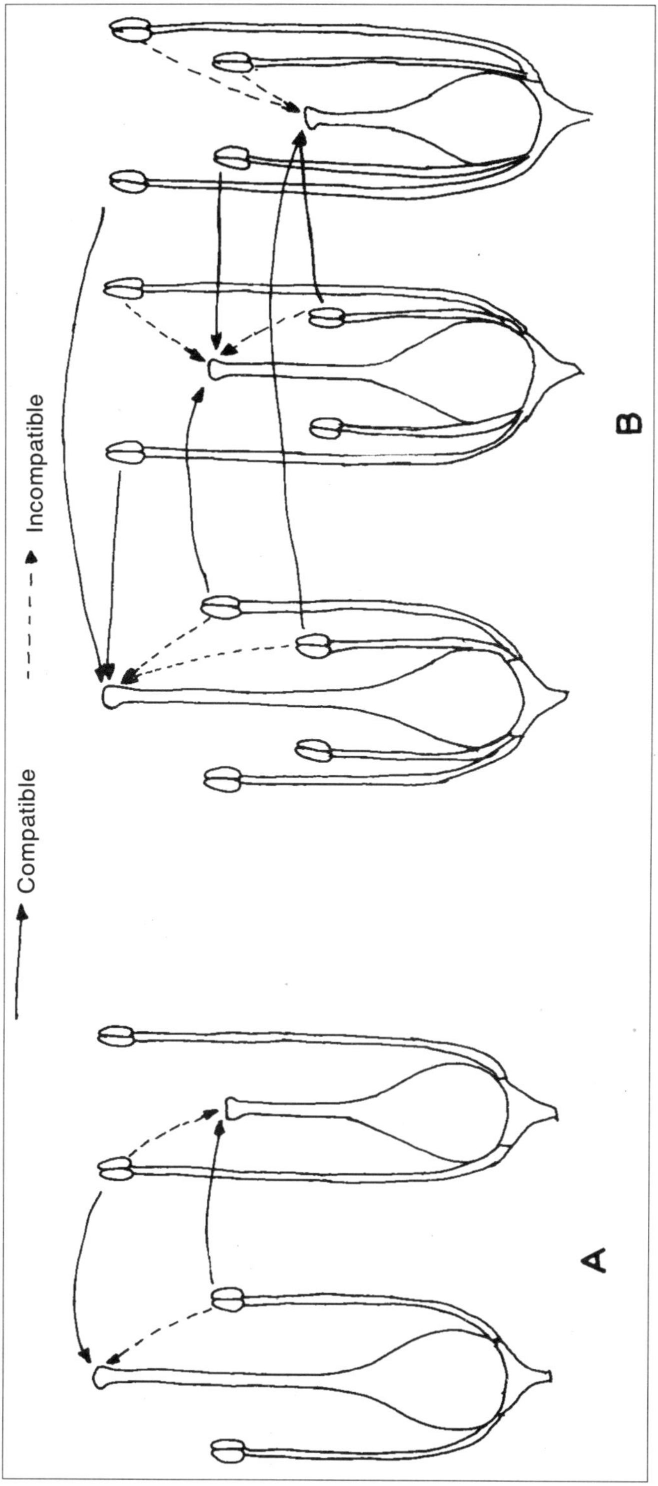

Figure 7.1
A: Distyly; B: Tristyly

Table 7.1: Heterostylous Families (adapted from Vuilleumier, 1967)

Family	*Type of Heterostyly*
Polygonaceae	D
Capparaceae	D
Saxifragaceae	D
Connaraceae	D
Leguminosae	D
Oxalidaceae	D, T
Linaceae	DE
Erythroxylaceae	D
Guttiferae	D
Turneraceae	D
Lythraceae	D, T
Primulaceae	D
Plumbaginaceae	D
Oleaceae	D
Loganiaceae	D
Gentianaceae	D
Polemoniaceae	D
Boraginaceae	D
Rubiaceae	D, T
Commelinaceae	D
Pontederiaceae	D, T
Iridaceae	D
Amaryllidaceae	D

D: Dimorphic; T: Trimorphic

(b) Homomorphic Incompatibility

In homomorphic incompatibility the flowers produced by different plants do not show morphological variations. Homomorphic incompatibility has been reported in more than 250 genera belonging to atleast 71 families. In homomorphic cases, physiological barriers act in such a way that pollen grains do not germinate on a stigma of similar genetic constitution or their growth is so slow that by the time pollen tube reaches the embryo sac, the latter withers out.

Depending on the origin of factors determining the mating types on the pollen side, two categories of self-incompatibility have been recognised:

1. Gametophytic Self-Incompatibility (GSI)

The incompatibility process is controlled by the genotype of the male gametophyte (pollen) itself, *e.g.*, Solanaceae, Onagraceae, Rosaceae, Poaceae, Fabaceae,

Scrophulariaceae. In some species, *e.g., Phalaris, Physalis* etc., two loci (*S* and *Z*) govern incompatibility, while in some others, *e.g., Beta vulgaris* and *Papaver*, three loci are involved. In these cases, polyploidy does not affect the incompatibility reaction. Pollen tube grows very slowly in the style containing the same *S* allele as the pollen and fails to effect fertilization. Therefore, all the plants are heterozygous at the S locus.

2. Sporophytic Self-Incompatibility (SSI)

The incompatibility process is determined by the genotype of the sporophytic tissue of the plant from which the pollen is derived, *e.g.*, Asteraceae, Brassicaceae.

Genetic Basis of Self-Incompatibility

Genetic basis of self-incompatibility was for the first time proposed by East and Mangelsdorf (1925) in *Nicotiana sanderae.* According to them, incompatibility reactions are controlled by a single gene, called S-gene, which has several alleles - S_1, S_2, S_3, S_4 S_n. Each plant has only two of these alleles. Cells of styles and stigmas are diploid and carry both the allels of the plant. Pollen grains are haploid and, therefore, carry only one of the two alleles. A pollen grain carrying a self-sterility allele does not grow well on a female style carrying the same allele, but grows and fertilises a plant that carries other alleles. A successful fertilization always denotes the presence of an allele in the pollen that is different from two alleles of the female plant. For example, a plant with the genotype S_1, S_2 produces two types of pollen grains, S_1 and S_2, none of which grows on a style with these alleles. Thus self-fertilization is ruled out. But if a plant with genotype S_2, S_3 is pollinated by such pollen grains, only the S_1 pollen grains germinates and fertilises whereas S_2 pollen grains fail to grow. On the other hand, all the pollen grains of a $S_1 S_2$ plant are viable and active on a $S_3 S_4$ tissue (Figure 7.2). Such a system leads to the production of heterozygous progeny, *e.g. Trifolium, Nicotiana, Lycopersicon, Solanum, Petunia* etc.

It should be emphasized here that in all the situations described so far it is the *S*-allele of the pollen or the male gametophyte which determines the incompatibility reaction (GSI).

In the Sporophytic Self-Incompatibility (SSI) also, the self-incompatibility is governed by a single gene, S, with multiple alleles; more than 30 alleles are known in *Brassica oleracea.* In general, the number of S alleles is considerably large in the gametophytic than in the sporophytic system. The incompatibility reaction of pollen is governed by the genotype of the plant on which the pollen is produced and not by the genotype of the pollen. It was first reported by Hughes and Babcock in 1950 in *Crepis foetida* and by Gerstel in *Parthenium argentatum* in the same year. In the sporophytic system, the *S* allele may exhibit dominance, individual action (Codominance) or competition. Consequently there may be many complex incompatibility relationships. In Sporophytic Self-Incompatibility (SSI) systems all the pollen of a plant behave similarly, irrespective of the *S*-allele they carry. For example, from a plant carrying $S_1 S_2$ alleles the pollen carrying S_1 or S_2 allele behaves as S_1 if S_1 dominant, or S_2 if S_2 is dominant; if there is no dominence both will behave as S_1 plus S_2. In other words, the presence of even one of the alleles of the stylar tissue in the sporophytic tissue of the male would render all the pollen of that non-functional

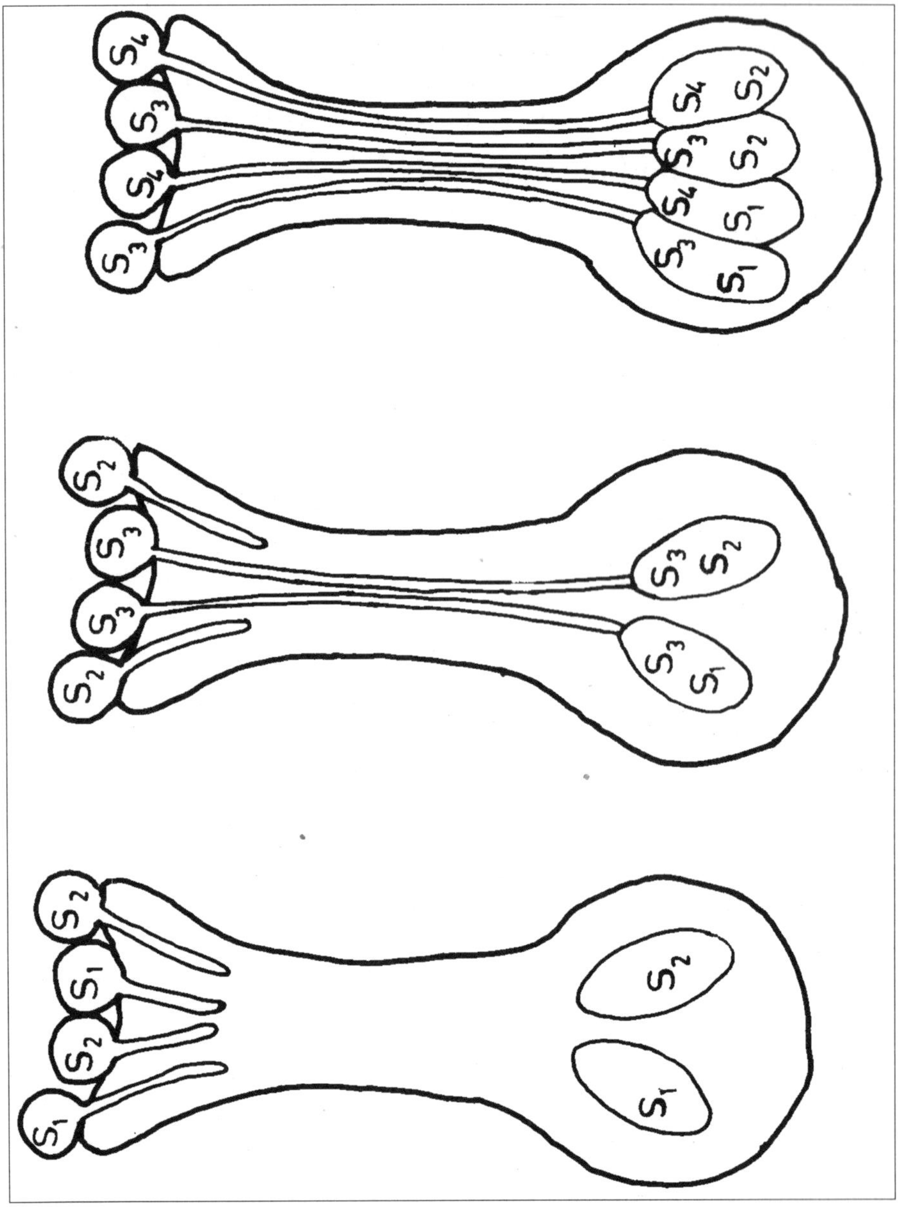

Figure 7.2: Gametophytic Self Incompatibility

with respect to that particular style. A $S_1 S_2$ plant, therefore, is completely incompatible to plants carrying $S_1 S_2$, $S_1 S_4$, $S_1 S_5$ or $S_2 S_3$, $S_2 S_4$, $S_2 S_5$ and so on, but shows 100 per cent compatibility with a plant carrying $S_3 S_4$, $S_3 S_5$ and so on (Figure 7.3).

For these two types of Incompatibility systems (GSI, SSI), Brewbaker (1957) observed a correlation between pollen nuclear number and the place of incompatibility inhibition in homomorphic forms.

a. Gametophytic

Observed primarily in species having binucleate pollen with inhibition occurring during pollen tube growth.

b. Sporophytic

Linked with trinucleate pollen grains and inhibition occurring on the stigmatic surface or very early tube growth.

Although this correlation holds good in a large number of taxa, there are many exceptions. For example, members of Poaceae are characterised by 3-celled pollen with gametophytic type of incompatibility.

Once an organism becomes evolutionarily adapted to a particular environment, or set of selective forces, it can usually maintain itself indefinitely in a state of ecological equilibrium with the other biotic elements to which it is adapted as long as the total environment itself remains essentially stable. Under such conditions, selection would be for continued stabilization of the organism; increased uniformity rather than increased variability would have the most selective value; homozygosity would tend to increase and heterozygosity would tend to decrease. In plants, which have evolved a much more series of reproductive methods than have animals, strong stabilizing selection might also lead to changes in the breeding system itself – there might be a shift to more inbreeding (self-pollination) and possibly to some form of apomixes or asexual reproduction.

Among obligate outbreeding plants it is possible that they might be highly heterozygous genetically even though they are quite uniform phenotypically. Also, heterozygosity itself may increase individual survival values and also population fitness through heterosis and homeostatic effects, while the increased genepool possible with heterozygosity provides a relatively high degree of evolutionary potential under changing environmental conditions or changes in the environmental selective forces that might be incurred by migration.

With enforced outbreeding, the accumulation of particular alleles, or groups of alleles, can most easily take place only in relatively small groups of interbeeding individuals that are in some way separated from other groups or populations of the same species with which they would normally interbreed. The organism belonging to separated, or allopatric, populations therefore tend to differentiate more readily from each other than do the sympatric organisms of a single population. If two or more populations of a particular organism are separated or isolated geographically for a long enough time (*i.e.*, for enough generations) it is possible that, in addition to any morphological differences that might develop, certain physiological or genetic

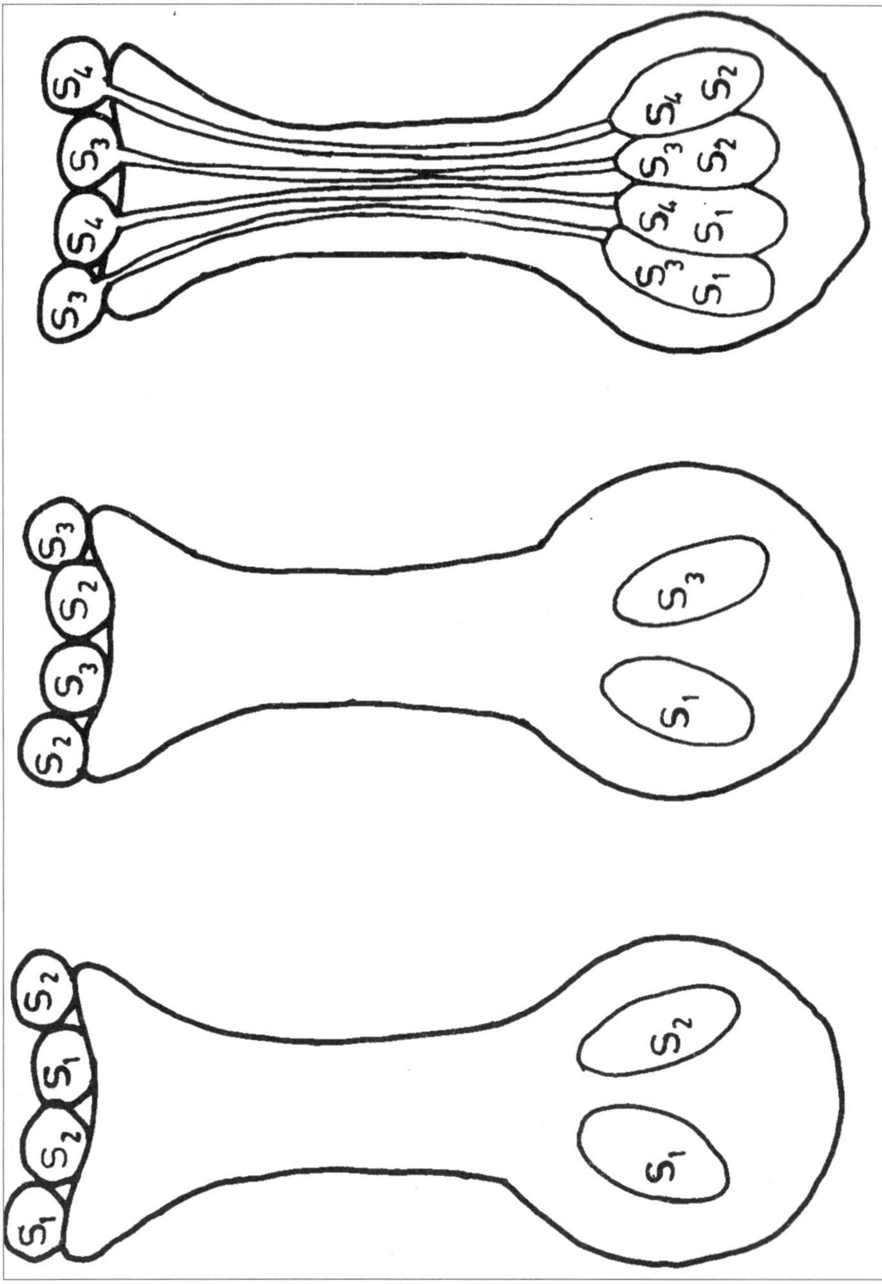

Figure 7.3: Sporophytic Self Incompatibility

changes would evolve that would prevent effective interbreeding between members of the two populations should they now again come into contact at some point. If this occurs, the two forms of the organism are said to be reproductively isolated and one of the first evolutionary steps toward the possible formation of a new species has been completed.

For many years, it was believed that because of the strong recombination effect of obligate outbreeding, geographic isolation was a necessary first step in the differentiation of races and the development of other more effective isolating mechanisms. On the other hand, there is now some experimental evidence to indicate that strong disruptive selection, even with maximum crossing or interbreeding of the two divergent forms of a polymorphic population, can be sufficient to maintain, and presumably isolate, two divergent group of individuals within the same population. Such sympatric speciation is most likely a rare occurrence in nature, but it should no longer be considered completely impossible. Given a particular combination of generation time, population size and polymorphism and a strong disruptive selective force, sympatric speciation could be realized.

Many of the higher plants are bisexual and capable of self-pollination and self-fertilization. If a single flower bears both male and female reproductive structures that is, both stamens and pistils, it is said to be a perfect flower. An imperfect flower would be unisexual, bearing either stamens or pistils but not both. If a single plant produces both male and female flowers, as is the case in many wind pollinated species such as maize (*Zea mays*), oak trees (*Quercus* sp.), and alder (*Alnus*), the plant is said to be monoecious. In a modification of the monoecious condition., staminate and perfect flowers may occur on one plant (andromonoecism) or pistillate and perfect flowers may be found on a plant (gynomonoecism). Heterozygosity and evolutionary potential are more strongly assured if the unisexual flowers are borne on separate plants. Such plants, as indicated previously, are said to be dioecious.

In the case of plants with perfect flowers, self-pollination is not only possible but is, in many cases, the primary type of pollination. Such closed and continued inbreeding tends to reduce recombination, reduce heterozygosity, reduce variability, and reduce evolutionary potential. On the other hand, self-pollination makes it possible for a single plant to reproduce and start a population; it can also insure adequate seed set in some plant populations in the absence of sufficient or appropriate specialized pollinators. In monoecious plants, self-pollination certainly possible but not probable, since the pollen is usually distributed by the wind and the female flowers of one plant are much more likely to receive pollen from the male flowers of neighbouring plant than from the male flowers on their own plant. Here, outbreeding is predominant, and recombination, heterozygosity, and evolutionary potential are accordingly increased.

Outcrossing is obviously obligatory for dioecious plants, and this raised several problems in terms of the optimum sexual composition of a population in terms of the ratio of male and female plants. Very often, the first problem is "solved" by the fact that, as previously mentioned, many dioecious plants do actually produce a few perfect flowers capable of self-pollination. Also, many of the species of dioecious plants are perennial; annual seed production is not essential to survival of the population. In the second case, selective pressures, possibly related to some aspect of conservation of energy within the population, have caused a shift to more female plants and fewer male plants in some species of dioeocious plants.

If the male and female parts of a perfect flower, or the male and female flowers of a given monoecious plant, mature a few days apart, the plants can-not self-pollinate and must outcross. If the pistillate flowers (or the pistilate portion of perfect flowers) mature first the plant (or flower) is said to be protogynous. In protandrous plant (or flower) the reverse is true; the stamens of a given flower mature and shed their pollen before the stigmas of the flower are receptive.

Strong protandry has apparently produced a shift toward the andromonoecious condition in several species of the carrot family (Apiaceae). Here the numerous small flowers are borne in aggregations called umbels. Large compound umbels, as in wild carrot (*Daucus carota*), are made up of many small ummbellets. If the individual perfect flowers are protandrous, the first flowers to open in the umbel may be pollinated by flowers nearer the centre of the umbel that open later. As this process continues, it is evident that the stigmas of the central flowers of the terminal umbel will be receptive after all of the pollen from the outer flowers of the lateral umbels (which begin to open as the central flowers of the terminal umbels finish blooming) furnish the pollen necessary to pollinate these central flowers of the primary umbel.

In a number of other plants, self-pollination is still possible, despite strong protandry. In many members of the Asteraceae, for example, the pollen is shed as the stigma, still morphologically and physiologically immature, pushes up through the ring of anthers. The mass of sticky pollen is literally pushed out of the flower, and most of it is usually soon picked up by the pollinator. As the stigma matures, it separates, exposing its pollen-free receptive inner surface. At this stage, pollen from another source, when placed on the stigma, would effect cross-pollination, However, if cross-pollination does not occur within a certain time, the tips of the stigma begin to reflex until the receptive inner surfaces come in contact with the style which may still be covered with pollen that was not picked up by the pollinator. In this way, self-pollination is accomplished and seed production assured even if cross-pollination has not taken place.

In some plants with perfect flowers self-pollination is unlikely because of the arrangement of the anthers and stigmas within the flower. If intraspecific differences in floral structure are responsible for the enforced out-crossing, the incompatibility system is said to be heteromorphic. Heterostyly is a good example of heteromorphic incompatibility and is found in *Oxalis* (Oxalidaceae), *Houstonia* (Rubiaceae) and a number of other genera. As the name implies, the flowers of one species are of two forms: some flowers have long styles and some flowers have short styles. In addition, the stamens of the long-styled (or pin morphs) flowers are well below the stigma while the stamens of the short-styled (or thrum morphs) flowers are above the stigma. In this way cross-pollination between "pin" and 'thrum" flowers is made more likely than self-pollination. As usual, the system is not absolute, and sometimes 'illegitimate cross-pollination' may occur; in this situation, pollen from the low stamens of one-long-styled flower reaches the stigma of another long-styled flower rather than the stigma of short-styled flower, or vice versa. However, even when illegitimate pollinations are made, fertilization may not occur because of secondary morphological or physiological adaptations correlated with heterostylous condition. 'Thrum' pollen is often larger than 'pin' pollen, the papillae on the stigmatic surfaces are different in the two flower forms, and there are often differential rates of growth of the pollen tubes, the best growth occurring when the pollen type is the opposite of the style type (*i.e.*, when the pollinations are 'legitimate). To function effectively as a unit, the various genes controlling all of the correlated characteristics associated with the two flower types, 'pin' and 'thrum' must be closely associated, perhaps as a supergene or block of genes, on a single chromosome. But even such linkage can be variously modified under different selective forces. If the survival value of self-pollination is great enough, heterostyly may be modified to homostyly, in which the styles are of constant length and the stamens are attached to the corolla tube in such a position that self-pollination is both possible and probable. In another modification, tristyly, the anthers of a single flower are at two levels and there are three style lengths correlated with specific stamen positions.

In other plants, which produce perfect flowers of only one form in which self-pollination cannot be prevented, genetically controlled physiological factors may prevent self-fertilization. This condition, called homomorphic self-incompatibility, effectively blocks inbreeding and may be either of two types: gametophytic self-incompatibility in which the inhibition of fertilization is the result of gene action in the pollen grain (the male gametophyte) itself as the pollen tube grows through the style; the sporophytic self-incompatibility in which the inhibition of pollen germination or pollen tube growth is imposed by the gene action of the sporophytic tissue of the stigma. In those cases that have been investigated, there is a correlation between binucleate pollen and gametophytic self-incompatibility and the formation of trinucleate pollen and the sporophytic type of self-incompatibility. In both forms, however, only one or two genes, and their alleles, seem to be involved. Although found in about a dozen unrelated plant families, the genetic basis of the self-sterility mechanism is remarkably similar in all the cases: the pollen tube will not grow on a stigma or through a style if the pollen and the stigma or style have a particular allele in common. Thus, a plant of the genotype S_1S_2 would produce two kinds of pollen, S_1 and S_2, neither of which would grow through the S_1S_2 sporophytic tissue if the pistil of the flower of the plant that produced the pollen. Such pollen grains would, however, be effective on a stigma or in a style with another genotype such as S_3S_4 or S_4S_5.

In the ground cherry, *Physalis* (Solanaceae), self-incompatibility is due to two independently assorting genes, S and Z and their alleles. This imposes an even more stringent selection for outcrossing: in the cross between two plants of the genotypes $S_4S_5Z_1Z_6$ and $S_1S_4Z_2Z_6$ for example, eight genetically different types of pollen would be produced but only ¼ of the pollen produced by plants of the first genotype would be effective on the stigmas of the plants with the other genotype. The wasted pollen is part of the biological 'cost' of maintaining heterozygosity and a greater evolutionary potential in these particular plants.

Although sexual reproduction is the mechanism which insures variability and evolutionary potential through recombination, asexual reproduction has strong survival value under certain conditions in which the production of uniform (but not necessarily homozygous!) preadapted genotypes can maintain, or even increase, a population in a given environment. Or, in cases where various forms of sterility prevent effective sexual reproduction, asexual reproduction is obviously the only method of survival of the organism, and in the higher plants several secondary types of asexual reproduction have evolved that provide for reproduction without the fusion of male and female gametes.

II. Asexual Reproduction

Vegetative propagation or reproduction is quite common among the perennial flowering plants and can take place by the formation of new plants from rhizomes, tubers, stolons and root divisions which become separated from the parent plant. Layering the rooting of detached branches, propagation from individual bulb scales, and the formation of vegetative propagules of various kinds on the leaves, stems, or in the inflorescences of certain plants can also produce large numbers of genetically identical individuals called a clone. If environmental conditions remain stable, the members of a clone may maintain a population for many years. Some clones, such as the long-lived coast redwood (*Sequoia sempervirens*) which reproduces by stump sprouts from old root crowns, and some of the rhizomatous grasses of the American prairies, provide examples of clones that have survived for thousands of years.

Because of uniform genetic background of the members of a clone, any variability between the members, even within a particular population, will probably be phenotypic variation caused by slight environmental differences. Clone material, therefore, is ideal for studying the morphological and physiological effects of different environments on a particular plant or series of plants from one or more populations from different areas of habitats.

Apomixis

Another form of asexual reproduction, called agamospermy, involves the production of viable seeds which have been formed without fertilization, although in some cases pollination is required to trigger the agamospermous process. The embryo in such seeds may develop from an unreduced egg (this is parthenogenesis) or in some other unreduced cell in the embryo sac, or it may develop from a somatic cell in the ovule. In all cases, the embryo is genetically identical with the seed-producing parent. If there are several different genotypes in the population, these same genotypes will be represented unchanged in the next generation, but possibly in different proportions. Although recombination is not possible, the series of populations maintains a small amount of flexibility through time as selection operates to increase or decrease the percentage of each genotype in each generation as conditions change. Such a situation occurs in the common dandelion (*Taraxacum officinale*) and numerous other, though usually perennial, members of the aster family (Asteraceae), grass family (Poaceae), and rose family (Rosaceae).

However, despite the immediate fitness exhibited by some asexually reproducing organisms, unless some sexual reproduction is possible, the members of a clone, as are most other organisms that result from asexual reproductive processes, are at an evolutionary dead end.

In those organisms that combine some degree of sexual reproduction, which insures recombination and evolutionary potential, with various forms of asexual reproduction, which provides for the rapid multiplication of adapted genotypes, we see excellent examples of evolutionary buffering. Such organisms have generally had long evolutionary histories and will most likely continue to survive under whatever drastic local or world-wide environmental changes that man causes.

III. Inbreeding Systems

Despite the long-term evolutionary advantages of cross-fertilization, a high degree of homozygosity and a positive pollination mechanism that regularly insures optimal seed production, even in the absence of pollinators, is of considerable selective advantage in some plant populations such as those of many annual plants or cloning species in new or ephemeral habitats. Here the advantages of immediate fitness (*i.e.*, survival) and the rapid production of numerous offspring preadapted to the environmental conditions of the new or temporary habitat far outweigh the long-term benefits to be derived from out-crossing. It is not surprising that many of our weedy annuals, under the strong selection for immediate fitness, are found to be autogamous, that is, they are self-pollinating, self-fertile and inbreeders.

The value of autogamy in either annual or perennial plants of colonizing species is also obvious. A single seed of a self-pollinating plant is capable of giving rise, in several generations, to an entire colony of plants. In many instances of dispersal, even long distance dispersal, it is not unrealistic to assume multiple introductions of a species into a given area over a period of time. The establishment of pioneer populations from different seed sources provides for some genetic variability in these populations that are disjunct from the main area of distribution of the species.

Since autogamy occurs in several unrelated taxa that are predominantly outcrossing, and since self-pollination is more likely to be present in disjunct populations of a given species, autogamy is considered to be an advanced or derived condition that has arisen independently in different groups of plants. The strong correlation between self-fertilization and colonization has been referred to as 'Baker's Law'.

A special situation has been described for several plant species which have dimorphic flowers and are regularly cross-pollinated near their centre of distribution but which have homomorphic self-pollinated flowers in the peripheral portions of their ranges. This occurrence of populations of self-pollinating plants in an otherwise outcrossing species can sometimes be correlated with environmental factors that greatly reduce the number of pollinators. The weather of the Faeroe Islands, in the north Atlantic off the coast of Norway, is often cloudy and there is much rain. As a consequence the number of insects is quite low and several widespread species of plants that are normally cross-pollinated by insects in continental Europe are adapted to self-pollination (by raindrops!) in the Faeroes. Autogamy is also frequent in plants that inhabit desert or semi-desert areas where flowering may be periodic and pollinators may be scarce. Other normally outcrossing plants may adapt to occasional unfavourable seasons or periodic fluctuations of temperature or moisture by what amounts to facultative autogamy. Under unfavourable environmental conditions only self-pollinated flowers are produced but under more nearly optimum conditions cross-pollination takes place. An interesting extension of this shift from self-pollination to cross-pollination in an individual in a single season can be seen in *Myosorus* sp. (Ranunculaceae) in which the flowers are normally self-pollinated. Here the carpels are separate on an elongated receptacle in the centre of the flower and the carpels present when the flower opens are pollinated with pollen from the same flower. If the environmental conditions remain static or get worse (*e.g.*, if the habitat gets too dry) no further carpels are produced. However, if the plants remain healthy and have abundant water the receptacles may continue to elongate and produce additional carpels above the already developing achenes formed earlier. The newer carpels will necessarily be cross-pollinated since no more pollen is available from that particular flower.

When we find both inbreeding and outcrossing in two closely related species, in a single species, or even in a single plant, it is quite likely that some floral dimorphism will be correlated with the two pollination types. Under reduced pollinator selection the self-pollinated flowers may be smaller, less showy, and have smaller anthers. The ultimate in this type of dimorphism occurs in those plants, such as *Commelina benghalensis, Viola papilionacea* and *Lamium amplexicaule,* which normally produce both open, showy, cross pollinated chasmogamous flowers and on the same plant also produce small, often drab cleistogamous flowers that never open but self-pollinated within the closed floral envelope. The cleistogamous flowers of *Lamium amplexicaule* (Lamiaceae), a small winter annual weed of lawns and waste places, are usually produced in the late winter or early spring and are follwed shortly by, or even intermixed with, the more showy chasmogamous flowers. In *Viola papilionacea* (Violaceaee), an herbaceous perennial or woodland clearings and open areas, the fragrant and showy chasmogamous flowers appear early in the spring when the days are short and the cleistogamous flowers follow with the longer days of summer as the new crop of leaves mature. Such plants as *Viola papilionacea* and *Lamium amplexicaule* thus have very effective mechanism to insure both outbreeding and inbreeding as a normal part of each reproductive cycle.

However, even in species that are predominantly self-pollinated, there usually remains the capacity for occasional outcrossing and a renewal of heterozygosity. A series of population studies of predominantly self-pollinated species of cultivated plants such as wheat, oats, and lima beans, has shown that normal outcrossing in these closely inbred plants may range from 1-12 per cent. In addition, the heterozygotes appear to be more vigorous and fertile than the homozygotes. This further indicates that self-pollination is a derived type of reproductive process that, under certain current conditions has a higher selective value than outcrossing despite demonstrated heterosis.

Consquences of Hybdridization

The existence of hybrids between two species can cause practical taxonomic problems, because such plants are not identifiable with any one species. If they remain infrequent and occur only where the two parental species meet they are nearly always readily recognizable as hybrids and the problems which they cause are minimal, especially when they are highly sterile. But in many cases they are fertile, and produce F_2 and backcross generations, which become common and widespread. Indeed, they can become commoner than either parent, for example various combinations in *Quercus, Betula, Ulmus* and *Crategus.*

Fertile hybrids may give rise to hybrid swarms, where by back-crossing and the production of F_2 and later generations the parental species become connected phenetically by every possible intermediate type, so that one species grades almost imperceptibly into the other. The parental limits can sometimes be detected by studying their variation in areas where only one of them occurs, or where they both occur but do not hybridize.

The existence of hybrid swarms indicates that there is a spectrum of ecological niches available to satisfy the requirements of a wide range of hybrid offspring. It is commonly observed that hybrid swarms (in fact hybrids in general) are particularly common in recently disturbed habitats, in which many new micro-niches have been artificially but unwittingly established. This was termed by Anderson 'hybridization of the habitat'. Often, however, this is not so, and hybrid swarms are not able to develop: introgression (or introgressive hybridization) may take place in such situations. This term was introduced by Anderson and Hubricht to describe the repeated backcrossing of a hybrid to one or the other parent, the hybrid product coming to resemble that parent quite closely after even a few generations but differing from it by some characteristics of the other species. Anderson and Hubricht spoke of 'an infiltration of the germplasm of one species into that of another. On the grounds that the ecological requirements of a hybrid phenetically closely resembling one parent are most likely also to resemble the requirements of that parent, introgression becomes a more likely outcome than the formation of hybrid swarms in situations where there are habitats suitable for the parents but not for intermediates.

Within a limited area introgression usually takes place only from one species to the other, but there are many examples known (*e.g., Quercus petraea* X *Q. robur*) where introgression occurs in one direction in some areas and in the opposite direction in others. Moreover, the same pair of species can in other situations give rise to hybrid swarms. It seems to largely the effect of environmental factors which decide whether hybrid swarms or introgression (and the direction of the latter) result. These factors are not simply the presence of different ranges of ecological niches, but include the habitat or pollen-source preferences of pollen vectors, and the sexual direction of the original cross (hybrids are likely to occur close to and therefore more likely to backcross with their female parent).

Introgression often occurs when the F_1 hybrid is relatively infertile, presumably because it is then much more likely to backcross to a fully fertile parent than to mate with a second infertile F_1 hybrid to produce an F_2 hybrid. F_1 hybrid which are rare are also more likely to back cross than to produce and F_2 generation. Indeed F_1 hybrid which are rare, ill-adapted to their environment and largely sterile appear to be very effective bridges for interspecific gene-flow. Moreover, the products of hybridization are likely to be most successful in evolutionary terms when they closely resemble one or other parent, as they then simply act as a means of enriching the gene-pool of an already successful species rather than producing a totally new sort of individual.

In some areas hybridization between two species has become so prevalent that the majority of individuals are of hybrid origin, and in extreme cases 'pure' examples of one or even both species no longer occur in that locality. One often reads or hears of species being 'hybridized out of coexistence'. Hybridization probably does sometimes assume such an aggressive role, but hybridization by long-range dispersal of pollen, or extinction of one of the parents for reasons not related to hybridization, can produce the same end results.

Stabilization of Hybrid

Although the occurrence of hybridization often obscures the distinction between species, in other cases it gives rise to new entities which are treated as separate species. This topic is often considered under the heading of the stabilization of hybrid progeny. It can arise in a number of ways of which Grant enumerated seven: (1) vegetative propagation, (2) agamospermy, (3) translocation heterozygosity, (4) unbalanced polyploidy, (5) amphidiploidy, (6) recombinational speciation, and (7) hybrid speciation. Grant discussed these seven phenomena under five headings: the clonal complex (1), the agamic complex (2), the heterogamic complex (3 and 4), the polyploidy complex (5) and the homoploid complex (6 and 7).

The most well-understod of these is amphidiploidy (polyploidy complex). It involves the formation of a new fertile taxon, which behaves as a diploid, by the doubling of the chromosome number of a sterile interspecific hybrid. If one includes all the diploidized segmental allopolyploids, it seems likely that between 20 per cent and 50 per cent of all vascular plant species arose in this way.

Simple F_1 hybrids have often become stabilized in various ways so that they have been regarded by taxonomists as species. Stace (1975) considered this treatment advisable whenever hybrids 'have developed a distributional, morphological or genetical set of characters which is no longer strictly related to that of their parents'. In other words, if the hybrid has become an independent, recognizable, self-reproducing unit, it is *de facto* a separate species. Totally or highly sterile hybrids have often become extremely successful species by adopting either a vigorous mode or vegetative or asexual reproduction or an agamospermaous mode of pseudo-sexual reproduction. These are the clonal complex and agamic complex respectively of Grant.

Under his heading homoploid complex Grant placed all those mechanisms which involved no loss of sexuality and no change of ploidy level or genetic mechanism. For example, in the genus *Gilia*, there are a number of well-defined diploid species of which some (*e.g.*, *G. achilleaefolia*) seem undoubtedly to be of hybrid origin. But in this genus there are a number of other 'hybrid species' which have become tetraploids (*e.g.*, *G. clivorum*), and in most examples of homoploid complexes polyploidy seems also to play a part. In *Euphrasia*, Yeo (1975, 1978) recognized three consequences of hybridization: hybrid swarms, incipient speciation and introgression, of which the second is relevant here. Incipient speciation in this genus involves the formation of 'extensive populations of comparatively uniform plants apparently of hybrid origin, usually a diploid and a tetraploid giving rise (via a triploid which produces some haploid gametes) to a new diploid, although some examples all at the diploid level are known. Ubsdell (1979) discovered in Lancashire a new fertile allohexaploid species, *Centaurium intermedium*, which she believed arose by the backcrossing to one of its parents of a largely infertile tetraploid hybrid between two tetraploid species. Presumably the hexaploid level arose due to the formation of some unreduced (tetraploid) gametes.

A number of examples of species and lower taxa which have arisen as introgressed variants of established species have been described, mostly in North America. One of the most striking is the postulated origin of *Purshia glandulosa* from the introgression of *Cowania stansburiana* into *Purshia tridentata*, involving two separate genera of woody Rosaceae (Stebbins, 1959).

Grant's heterogamic complex is represented by two relatively unusual mechanisms by which hybrid plants can discover a successful formula for adaptation and survival: translocation heterozygosity and unbalanced polyploidy. The best example of the first are found in many species of *Oenothera* which form multivalent rings of chromosomes at meiosis, and of the female parent contributing four sets and the male parent only one set of chromosomes to the zygote.

There are good grounds for considering hybridization as one of the major evolutionary mechanisms in plants. It seems plausible that a phase of hybridization is followed by one of stabilization, and that this pattern is repeated cyclically (Ehrendorfer, 1959). Hybridization represents a period of maximum recombination, and stabilization the selection of the fittest recombinants so formed. Fortunately for botanists, different groups of plants occupy different parts of this cycle at any one period, so we are able to study all aspects of it at the present time.

Ideal Species

As a starting point one may take the example of an 'ideal species'. By this is meant a taxonomically ideal species, *i.e.*, one which poses no taxonomic problems, as it is always recognizable as a distinct entity. Thus it has no sharp discontinuities of phenotype within its spectrum of variation, and does not merge with other species.

In most instances such species are genetically isolated from (*i.e.*, unable to interbreed with) other species and are at least partially outbreeding. These are the 'sexual, outbreeding, non-hybridizing species' discussed by Stace (1978). Species of this type are very common in, for example, many Fabaceae and Apiaceae, *Sedum*, *Campanula* and *Allium*, although there are exceptions in all these examples.

Most (though not all) taxa which pose intrinsic taxonomic problems do not correspond with the above situation. Instead, either they hybridize with other taxa so that the genetical limits are wider than the morphological ones, or they exhibit breeding barriers between different elements of a morphologically recognizable taxon. These two categories are discussed separately below.

Hybridizing Species

The subject of plant hybrids and hybridization has been treated in detail by Stace (1975) with special reference to the British flora. In this chapter only the salient points relating to taxonomy are covered.

A hybrid has been defined as 'a zygote produced by the union of dissimilar gametes' (Darlington, 1937). Such an extremely broad definition is quite valid in a genetical sense, but is useless in a taxonomic one since it covers almost all animal individuals and the majority of plant individuals. A more useful definition is of a taxonomic hybrid – the product of breeding between distinct taxa. One can be more or less specific by limiting the context to taxa of particular levels, for example interspecific hybrid and intergeneric hybrid. Usually taxonomists, when speaking of hybrids, are referring to interspecific (or intergeneric) hybrids rather than to hybrids between taxa within one species.

The Extent of Hybridization

The notion that interspecific hybrids are rare is ill-founded. The only modern broad compilations of their numbers (*i.e.*, number of species combinations) are those of Knobloch (1972, 1976) who recorded 23675 in the angiosperms and gymnosperms and 620 in the pteridophytes. However this was an uncritical survey, many fictitious hybrids being included, as were those only artificially produced. In terms of well-substantiated, naturally occurring hybrids the figure is much lower, but there must be great many undetected hybrids in underworked areas, especially the tropics. Realistic figures can be obtained only from very well studied floras, such as that of the British Isles. Stace (1975) recorded 626 well substantiated interspecific hybrids among vascular plants, plus 122 possibly correctly identified ones (as well as 227 errors) in the British Isles; since that compilation at least 16 additional combinations have been discovered. Taking into account the doubtful cases above, it is reasonable to assume that there are about 700 interspecific hybrids among the British vascular plant flora of about 2500 native and alien sexual species (excluding apomictic taxa), ignoring the many combinations recorded from only outside the British Isles yet involving two British species. The figure of 2500 represents, very approximately, 1 per cent of the world's vascular flora, which might therefore include in the region of 70,000 different naturally occurring interspecific hybrids. Although most of these are uncommon, clearly hybridization is not a rare or abnormal phenomenon. Moreover, a great many firmly accepted species are known to be of hybrid origin. When one considers the potential for the hybridization, hybrids assume even greater importance. There are vastly more artificial hybrids known that natural ones; in the Orchidaceae alone 45,000 have been synthesized. Indeed, one is forced to the conclusion that in general the ability to hybridize is the usual situation (Raven, 1976). In the British flora about half of the genera are monotypic, and about half the rest form at least one hybrid in the wild.

A small proportion of naturally occurring hybrids involve three or even more species. This, of course, is always a possibility when a primary (dispecific) hybrid is fertile, since it might then mate with a third species. About 26 such hybrids have been recorded from the British Isles. Again, the possibilities are much greater in the case of artificial hybridizations, and a hybrid incorporating 13 different parental species of *Salix* has been synthesized in Sweden (Nilsson, 1954).

In the non-vascular groups much less is known concerning natural hybridization. In terms of recorded numbers hybrids are much less common than in the vascular plants, but probably most have been overlooked. In the bryophytes they are usually sterile, so they exist only as hybrid sporophytes upon a female gametophyte and are relatively difficult to detect. The fact that they have been recorded in a quite wide range of bryophytes, and also of algae, suggests that hybridization is an important phenomenon in all groups of plants (as well as in fungi).

Intergeneric hybrids are much less common. In the British vascular plant flora there are only 29 interspecific combinations of intergeneric hybrids known, involving only 14 intergeneric combinations. Extrapolated to a world scale, this indicates that there might be about 250 intergeneric hybrid combinations. But again, there are many more artificially produced intergeneric hybrids, and natural intergeneric hybrids occur in algae and bryophytes as well as in the vascular plants. 'Wide' hybridization is particularly a feature of the Orchidaceae, where artificial hybrids involving as many as five different genera in one offspring have been synthesized. In the Poaceae, another family in which intergeneric hybrids are rather frequent (although here, unlike the Orchidaceae, they are mostly sterile), all naturally occurring hybrids are between genera within one tribe, although intertribal crosses have been artificially synthesized. On the other hand no hybrids have so far been discovered or synthesized between plant species assigned to different families.

Little generalization can be made concerning the frequency of hybridization since between extreme cases there is every intermediate. At one extreme one may cite the case of the grass *Desmazeria marina* X *D. rigida*, known only as a single plant found in 1960 in Wales (Benoit, 1961). Both parents and the hybrid are annuals, so that this individual lived only for a few months. At the other extreme come hybrids such as *Geum rivale* X *G. urbanum*, *Betula pendula* X *B. pubescens* and *Circaea alpina* X *C.lutetiana*, which appear to occur whenever the two parents come into contact, and which in many regions have become more common than one or even both parents. Naturally, the majority of hybrid combinations come somewhere between these extremes, and most nearer the former situation, but there are certainly a great many very common and widespread hybrids.

Chapter 8
Methods in Experimental Taxonomy

The techniques of experimental taxonomy comprise methods of testing to determine to which of the above-described biosystematic units a population belongs. The most reliable evidence and conclusions are obtained from application of all methods: those of orthodox taxonomy as well as those of cytology and genetics, combined with cultivation in uniform and in varied environments. Ordinarily the methods of testing in uniform and in contrast environment are applied simultaneously to a number of populations of the same and of different taxonomic rank within a genus.

1. *Growth in uniform environment* provides a means of studying variability in heredity.

   Population samples of the same taxonomic category (*e.g.*, spcies, subspecies) are procured from different environments, and are grown in a common experimental plot or under controlled environmental conditions.This permits comparison of behavior of plants of unlike heredity in the uniform environment, thereby distinguishing between hereditary variation and environmental modification.
2. *Growth in varied environments*, as afforded by a series of field stations or artificially controlled environments, is of equal importance as a means of observing the interplay between heredity and environment.
3. Cytogenetic analysis may involve the 3 following types of techniques. (a) Cytological studies are made of as many populations as practical to determine the possible correlation's between visible chromosome

differences and differences in external morphology and geographic distribution within the species complex. (b) selected forms are crossed and the hybrid progeny grown to determine the fertility of F_1 generation, the fertility and vigor of F_2 generation, and to analyze the genotypes of related ecotypes and ecospecies. (c) The chromosomal homologies of these natural units are analyzed through studies of the chromosome pairing in their hybrids.

4. *Crossing programmes to test the presence or absence of sterility barriers* are a part of biosystematic study. The crossing test is applied to any taxa suspected of possessing these barriers, and in instances where the barrier is found not to be absolute the F_1 and F_2 generations must be studied statistically to determine their vigor and fertility. Such tests are conducted also with plants of populations presumed to be of close affinity to determine the genetic mechanism that may be allowing them to retain distinctness. The crossing tests are applied also to geographical or ecological extremes to determine if such extremes are separated by genetic barriers.

Isozymes, Allozymes

The taxonomic applications of the results of electrophoresis have been useful mainly at and below the generic level. While much work has been directed at food-reserve proteins (*e.g.*, in legumes and cereals), more has involved enzymes. Not only can electrophoresis separate closely related storage proteins and enzymes, but it has also led to the recognition of allozymes and isozymes, which are different forms of what were previously considered single enzymes. Allozymes are different forms of an enzyme where the constituent polypeptides are determined by different alleles at one locus; isozymes (or isoenzymes) are different forms where the polypeptides are determined by more than one locus. The former appear to be the more common, but both are loosely termed isozymes in much of the literature. These isozymes or allozymes can be distinguished by their different electrophoretic mobilities and are thought to differ also in the precise conditions needed for their optimal catalytic activity; hence they might be brought into operation by the plant in different organs, at different stages of growth, or indifferent habitats. Enzyme polymorphism is often exhibited below the species level and especially when the constitution of more than one enzyme is worked out, isozymes can be used to recognize even individual clones. They are therefore of use not only to taxonomists, but also to ecologists and population biologists. In many cases inter-populational variation in isozymes is not accompanied by morphological differentiation, for example enzyme races of *Concocephalum conicum* in Poland.

The most valuable electrophoretic results have perhaps been obtained in the cereals related to wheat, in relation to the genomic constitution and ancestry of the tetraploids and hexaploids. For example, Johnson (1972), working on storage proteins, concluded that the hexaploid bread-wheat (*Triticum aestivum*) did indeed contain a sum of the proteins possessed by the diploid species, which had been postulated on morphological and cytological evidence to be ancestral to it. Barber (1970) studying enzymatic proteins, found that certain polyploids possessed the isozymes of all their progenitors plus some new ones. These new hybrid isozymes are presumably oligomers incorporating polypetides in new combinations of genomes in the polyploids.

Chapter 9

Source of Taxonomic Characters

Most systems of classification of plants have been evolved on the basis of exomorphic characters placing particular reliance on the characters of the flower, the flower being considered more conservative than the vegetative organs. Even so the goal of reaching a truly natural system of classification has remained unfulfilled.

From the beginning of the twentieth century, with the development of technology, newer evidences became available to taxonomists. These newer branches like anatomy, embryology, palynology, cytology, phytochemistry and ultrastructure have been utilized in deducing the relationships of plants. Taxonomists in recent times like Cronquist, Takhtajan and Dahlgren have utilized these newer evidences in classifying angiosperms. In the following pages the utility of these evidences in solving taxonomical problems is discussed.

Anatomy in Relation to Taxonomy

The first attempt to use anatomy in plant classification was made by Bureau (1864). He used the anatomical characters for delimitation of taxa of various levels in the family Bignoniaceae. Solereder (1899) in *Systematische Anatomie der Dicotyledonae* emphasized the importance of xylem structure in plant classification. Bailey (1951) and Metcalf and Chalk (1950) have listed numerous anatomical features that can be made use in taxonomy. The *Anatomy of Dicotyledons* (Metcalfe and Chalk, 1950) and *Anatomy of Monocotyledons* (Metcalfe, 1960 onwards) are some of the books in this direction.

Epidermal Characters

Stomata

Leafy characters especially epidermal characters and stomatal characters are very much useful in taxonomy. The importance of epidermal characters was emphasized by Stace (1965) while the micromorphology (study of morphological characters with scanning electron microscope) has been emphasized by Dehgan (1980) and Barthlott (1981).

Van Cotthem (1970) divided the stomata into 8 types. They are:

Anomocytic (Ranunculaceous Type)

Stoma surrounded by a limited number of cells that are indistinguishable in size, shape or form those of remainder of epidermis (Figure 9.1a). *e.g.* Ranunculaceae, Malvaceae, Papaveraceae, *Tridax procumbens, Boerhavia diffusa, Cleome* spp.

Anisocytic (Cruciferous Type)

Stoma surrounded by three cells of which one is distinctly smaller than the other two (Figure 9.1b), *e.g.*, Cruciferae, Solanaceae.

Paracytic (Rubiaceous Type)

Stoma accompanied on either side by one or more subsidiary cells parallel to the long axis of the pore and guard cells (Figure 9.1c). *e.g.*: Rubiaceae, Magnoliaceae.

Diacytic (Caryophyllaceous Type)

Stoma enclosed by a pair of subsidiary cells whose common wall is at right angles to the guard cells (Figure 9.1d), *e.g.*: Caryophyllaceae, Acanthaceae.

Tetracytic

Four subsidiary cells are present, two lateral and terminal (Figure 9.1f), *e.g.*, Most families of Monocotyledons.

Actinocytic

Stoma surrounded by a circle of radially elongate subsidiary cells (Figure 9.1e), *e.g.*, Commelinaceae, Musaceae.

Cyclocytic

Stoma surrounded by four or more subsidiary cells which form a ring around each stomata (Figure 9.1g), *e.g.*, Tribe Laguncularieae of the family Combretaceae.

Hexacytic

Stoma accompanied by six subsidiary cells consisting of two lateral pairs parallel to the long axis of the pore and two polar (terminal) cells; the second lateral pair as long as the stomatal complex.

Besides the above, several other characters can be used in taxonomy. Presence of stomata on the upper side (epistomatic) or lower side (hypostomatic) or on both sides (amphistomatic) are also useful in taxonomy. For example the presence or absence of stomata on the upper surface of a leaf is a good diagnostic feature in the separation of *Alnus subcordata* and *A. orientalis* of Betulaceae, which are otherwise difficult to distinguish. In the former epistomatic condition prevails while in the latter, the hypostomatic. Similarly shape of the guard cells are also useful in plant classification. For example stoma with bean-shaped guard cells are met with Dicotyledons while stoma with dumbell-shaped guard cells are seen in Poaceae. In Cyperaceae rectangular guard cells are seen.

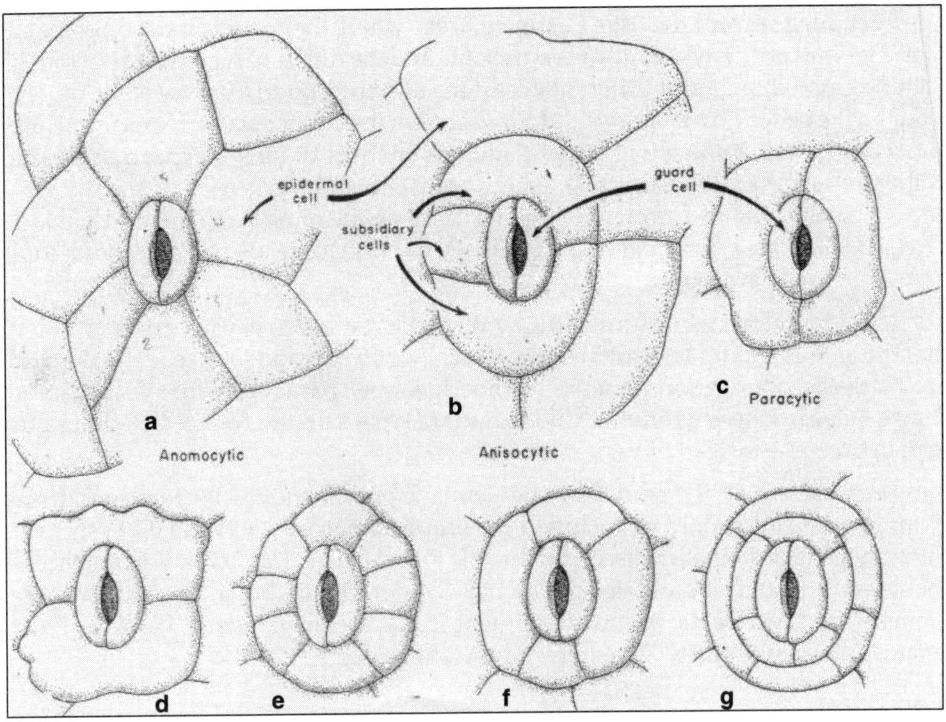

Figure 9.1a–g: Types of Stomata
a: Anomocytic; b: Anisocytic; c: Paracytic; d: Diacytic;
e: Actinocytic; f: Tetracytic; g: Cyclocytic

Development of stomata has been divided by Pant (1965) into three types.

Perigenous

Stoma in which the subsidiary cells do not have a common origin with the guard cells, but are formed by cells lying around the meristemoid that divides to form the guard cells.

Mesogenous

Stoma in which the subsidiary cells are produced from the same meristemoid (initial) as the guard cells.

Mesoperigenous

Stoma in which at least one of the subsidiary cells has a common origin with the guard cells, but others do not.

Development of stomata is useful in plant classification. Ramayya (1977) has emphasized the importance of stomatal characters in taxonomy. Ramayya and Rajagopal (1968, 1971) gave a key to genera of Portulacaceae and Aizoaceae based on epidermal characters.

In the family Acanthaceae the stomata are diacytic, whereas in the closely related Scrophulariaceae the stomata are anomocytic. In the family Combretaceae the stomata are paracytic in the subfamily Strephonematoideae and anomocytic in the subfamily

Combretoideae except the tribe Laguncularieae where they are cyclocytic. Stomatal characters are not, however always so reliable and the different types have obviously each originated on many different occasions and they often exist together on one plant. In the genera *Streptocarpus* and *Saintpaulia* of the Gesneriaceae; for example, the cotyledons bear aomocytic stomata and the mature organs anisocytic stomata (Saharsrabudhe and Stace, 1974). In *Lippia nodiflora* of Verbenaceae anomocytic, anisocytic, diacytic and paracytic stomata occur on the same leaf (Pant and Kidwai, 1964). In *Alternanathera pungens* of family Amaranthaceae also all the above four types of stomata are found.

Inamdar (1968) who studied the stomata of *Nyctanthus* expressed the opinion that this genus should be kept in Oleaceae. In *Nyctanthus* and Oleaceae the stomata are Tetra-Mesoperigenous (sometimes aperigenous), paracytic while Verbenaceae shows Dia-Mesogenous stomata. So Inamdar (1968) said the *Nyctanthus* should be kept in Oleaceae instead of Verbenaceae.

Bremekamp (1953) said that the genus *Elytraria* should be removed from Acanthaceae and should be included in Scrophulariaceae. Paliwal (1966) studied the phylogeny of stomata to ascertain this. He found that in *Elytraria* and Acanthaceae the stomata are diacytic and development is Syndetocheilic, but in Scrophulariaceae stomata are anomocytic and development is Haplocheilic. Hence Paliwal (1966) retained the genus *Elytraria* in the family Acanthaceae.

Trichomes

Trichomes are epidermal outgrowths. They exhibit great variation in structure and function and thus are of much taxonomic value. Glandular and non-glandular hairs, papillae, setae and prickles are all considered as hairs. These trichomes are useful in plant classification and understanding the relationship between groups. Schulz (1936) used the type of hair as the major criterion in the subdivision of the family Cruciferae into tribes and genera. Ramayya (1962, 1969) studied the trichomes of Compositae and gave key to the genera based on trichome characters. Based on trichome characters he considered Heliantheae as the primitive tribe. Rollins (1944, 1945, 1946, 1949) studied the trichomes of *Parthenium argentatum* and *P. incanum* and their natural hybrids. He has shown that in *P. argentatum* the trichomes are T-shaped while in *P.incanum* they are whip-like with a long thread. Intermediate types of trichomes are observed in the hybrids between these species. In the family Combretaceae Stace (1965, 1969 a, b, 1973) found that trichome anatomy is of immense significance in classification at all levels, from the circumscription of the family down to the separation of species and even varieties. In particular, it has led to an improved tribal classification within the family and an improved subgeneric and sectional classification within the largest genus *Combretum*.

Epidermal Cells

The epidermal cells, other than those modified by their relationship to trichomes, stomata, the venation system or other special structures, provide many characters of taxonomic value. These include their size, shape, orientation, anticlinal wall undulation and periclinal wall curvature. The epidermal cell modification occurs also on the leaf margin and when silicified, calcified, suberized, gelatinized or containing various types of crystals. Silicified and suberized epidermal cells are of major taxonomic importance in the Poaceae.

Other epidermal features are also useful in plant classification. The genus *Morus* is divided into species based on Idioblasts in epidermis. Tomlinson (1956, 1959 a, b) and Metcalfe (1968) said that silica bodies in the leaves are useful in classification of the families Zingiberaceae, Musaceae, Arecaceae and Rosaceae up to genera and even species.

Internal Structure

Carlquist (1961) said that leaves in Angiosperms show great diversity and this character is useful in classification. The nature of epidermis, its thickness, type of mesophyll, sclerenchyma patterns, sclereids, crystals, venation patterns, etc, may be useful in classification. The importance of foliar anatomy in the systematics of grasses was realized quite early by Duval Jouve (1875). According to Avdulov (1931) Poaceae have two types of internal structure in leaves. In the Festucoid (Pooideae) grasses the mestome around the stele is layered and mesophyll cells are straight while in Panicoid grasses mestome may or may not be present and the mesophyll cells are arranged radially.

Similarly Ayensu (1974) found that sclerenchyma patterns in the Velloziaceae can be used to separate the genera *Vellozia* and *Barbasenia*. The genera in Cyperaceae can be segregated based on internal structure of the leaves (Govindarajulu, 1969).

Crystals, especially calcium oxalate crystals, are found to be useful in the classification of Angiosperms. Foliar sclereids have, of late, been demonstrated to be of taxonomic importance in many taxa.

Laticifers are cells or a series of fused cells containing latex and are common features of many succulent plants and other plants of arid regions. They vary widely in their structure and latex in their composition. Their presence or absence and when present, their structure has been of some taxonomic value. In Aroideae (De Bary, 1884), for example, certain species lack laticifers or any related structures while others have longitudinal rows of elongated, cylindrical sac-like cells. Some authors consider that Fumariaceae, which lack laticifers (Sperlich, 1939), should be considered distinct from the related family Papaveraceae having a well developed latex system. In the former family laticifers are replaced by idioblasts. It is one of reasons for considering these two families as distinct form the rest of the families in Engler's Rhoedales.

The taxonomic usefulness of petiole vascularization patterns has become increasingly apparent in recent years (Radford *et al.,* 1974). Sinnot (1914) gave a classification of nodal types in megaphyllous plants. In general there are three major types of node – unilacunar, trilacunar and multilacunar (Figure 9.2). Trilacunar nodes occur in the majority of Dicotyledons, mutilacunar nodes in the primitive orders such as Magnoliales, Piperales, Trichodendrales and advanced orders such as Umbellales and Asterales and unilacunar nodes in Laurales, Caryophyllales, Ericales, Diapensiales, Ebenales, Primulales, Myrtales and a majority of families of Asteridae (Paliwal and Anand, 1978).

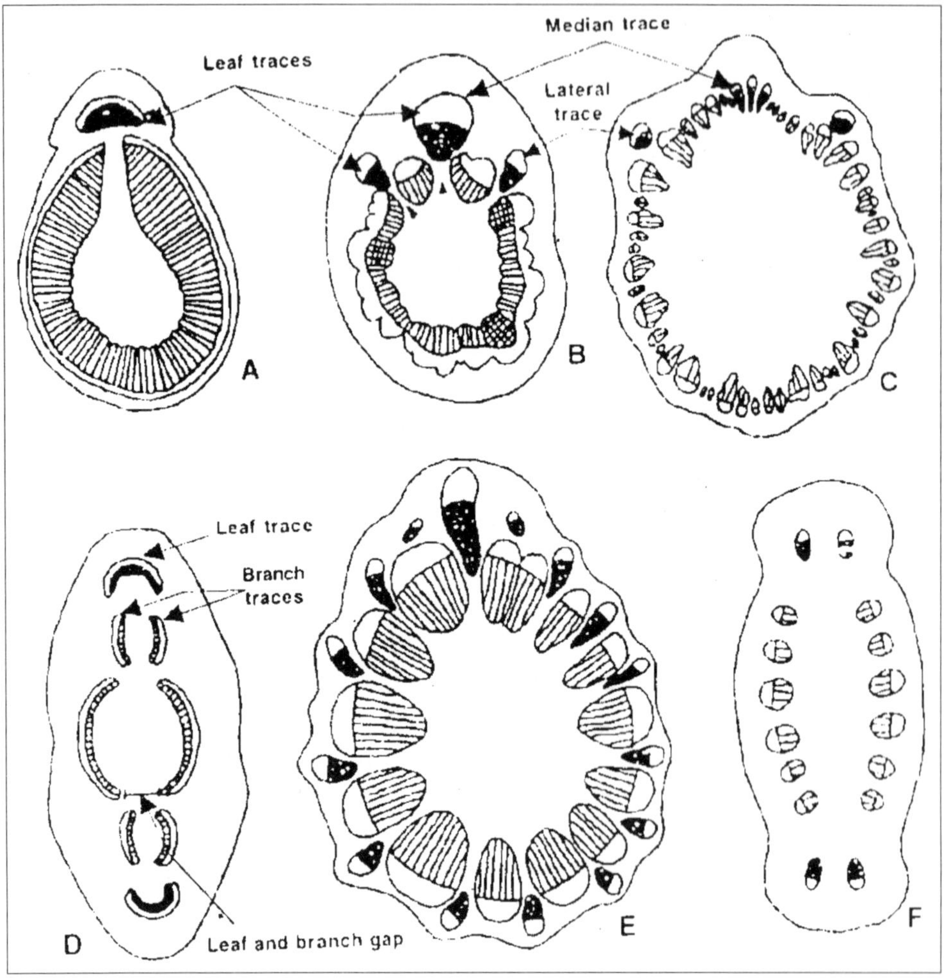

Figure 9.2: Types of Nodes
A, D and F: Unilacunar; B and C: Trilacunar; E: Multilacunar

Stem anatomy has been studied in the herbaceous members where it has proved to be of some diagnostic value. In the stem of *Ranunculus repens* there is relatively little strengthening tissue and is correlated with the creeping habit of the plant, while in the closely related, species, *R. acris,* interfascicular tissue is sclerified and therefore an erect habit is possible.

Stace (1970) has shown that in the subgenus *Genuini* of *Juncus,* anatomical characters of stem and foliar bract can be used to distinguish most British species. Further, it is also possible to identify parents of several hybrids on anatomical grounds.

There is considerable variety in the anatomy of stems of *Dioscorea* and can be successfully employed in the delimitation of species that are not easily separable on exomorphic grounds. Ayensu (1970) found anatomy useful in separating *Dioscorea rotunda* and *D. cayennensis,* which are closely related and difficult to distinguish.

The classical case of anatomy in relation to taxonomy is the distribution of vascular bundles. In Dicotyledons vascular bundles are arranged in a ring while in Monocotyledons vascular bundles are scattered. Bentham and Hooker (1862-83) included Basellaceae as a subfamily in Chenopodiaceae. Anatomically Basellaceae are characterized by a single ring vascular strands with intraxylary phloem in the larger bundles. On the other hand, Chenopodiaceae are characterised by the development of Polycyclic, anomalous secondary vascular bundles.

Bentham and Hooker (1873-1876) split the Gentianaceae into two subfamilies – Menyanthoideae and Gentianoideae. But Hutchinson (1969), Cronquist (1981) and Takhtajan (1980) raised them to independent families. Vascular bundles in Menyanthaceae are collateral while Gentianaceae show bicollateral vascular bundles.

The most important anatomical features that have been used in taxonomy and phylogeny are those of the secondary wood. This is because of their conservative nature. Wood anatomy can be used at all taxonomic levels. At and above the family level it has been successfully used along with other lines of evidence to establish the systematic position of the primitive vesselless Angiospermic families Winteraceae (all seven genera-*Drimys, Pseudowintera, Bubbia, Belliolum, Exospermum, Zygogynum* and *Tetrathalamus*), Trochodendraceae, Tetracentraceae, Amborellaceae and Chloranthaceae. Similarly it has been agreed by all phylogenists that the Englerian group of primitive Angiosperms, namely Amentiferae can not be considered primitive since they have specialized wood (Tippo, 1938). Anatomical evidence also supports the case for considering the remarkable genera *Paeonia* and *Austrobaileya* as the type of independent families (Bailey and Swamy, 1949).

At and below the generic level, the most useful contribution of wood anatomy is to provide evidence for placing of taxa of uncertain affinity. In *Quercus* (Fagaceae), different subgenera and sections can be delimited with the help of anatomical evidence. The placement of the genus *Myristica* and the related genera close to Lauraceae is supported by the evidence of wood structure (Eames, 1961), while on the same basis Myristicaceae are considered unrelated to Annonaceae and Eupomaticeae. Similarly, placing of Calycanthaceae in or near Rosales or Myrtales is not supported by wood structure.

Sieve cells also provide taxonomical evidence. For example, unspecialized sieve cells occur in lower vascular plants – Pteridophytes and Gymnosperms. Only one Angiosperm, *Austrobaileya scandens*, shows unspecialized sieve cells in phloem.

Floral Anatomy and Taxonomy

Floral anatomy has a wide application in determination of status of high ranking groups such as genera and species. On the basis of similarity of floral anatomical pattern affinities of families Papaveraceae, Capparaceae, Brassiaceae and Magnoliaceae could be established. These families have a common parietal placentation and inverted placental bundles with regard to floral axis on inner side of secondary marginal strands with normal orientation. This fact also supports placement of Moringaceae in Rhoedales. Puri (1947, 1949) suggested affinities between Passifloraceae, Moringaceae and Cucurbitaceae on floral anatomical basis.

Burtt (1965) opined that *Cyrtandromoea* has no close affinity with other members of Gesneriaceae and advocated its transfer to Scrophulariaceae with which it exhibits more resemblance. Singh and Jain (1978) have supported this treatment on the basis of floral anatomy. Kale and Pai (1979) have supported the removal of the genus *Trichopus*, on the basis of floral anatomy, from the family Dioscoreaceae of Bentham and Hooker and erection of a unigeneric family Trichopodiaceae as has previously been suggested on anatomical evidence by Ayensu (1966, 1972).

Fruit Wall and Seed Coat Anatomy

Fruit wall and seed-coat anatomy have also come to the help of taxonomists in many cases. The distinction between the two genera *Ammannia* and *Rotala* (Lythraceae) has been puzzling taxonomists for quite some time. The diagnostic feature was in the fruit-irregularly rupturing fruits in *Ammannia* and valvular capsules in *Rotala*. But this difference was very difficult to appreciate, especially in herbarium specimens. Van Leemjen (1971) found the fruit-wall character to distinguish the two taxa. The orientation of the cells of the two main layers of pericarp are different in *Ammannia* the cells of both layers are very similar, whereas in *Rotala* the cells of the inner layer are linear, strongly lignified and transverse to those of the outer.

Two different views were expressed regarding the systematic position of the Asteraceous genus *Lagascea*. Cassini (1815), Lessing (1832), DeCandolle (1836) and Merxmuller (1954) included the genus *Lagascea* in the tribe Vernonieae. Bentham (1873), Hoffmann (1894) and others place *Lagascea* in the tribe Heliantheae. Stuessy (1976) on the basis of morphological and chromosomal studies, supports the assignment of the genus *Lagascea* to Heliantheae. Pullaiah (1981) on the basis of pericarp structure favours retention of *Lagascea* in the tribe Heliantheae. In Heliantheae three different zones are distinguished in the pericarp. Between the middle and outer zones characteristic schizogenous cavities develop which get filled with a brown resinous and hardening substance. Such schizogenous cavities are absent in Vernonieae

Seed coat morphology has yielded valuable taxonomic informations. Seed coat anatomy has also been used for systematic purposes.

Ultrastructure in Relation to Taxonomy

Morphology and taxonomy have benefited greatly in recent years from information obtained by using advanced technology in microscopy, particularly Scanning Electron Microscopy (SEM) and Transmission Election Microscopy (TEM).

However, only a few ultrastructural characters are of wide application in plant classification. ' Dilated Cisternae' (DC) in the endoplasmic reticulum were first noted and described by Bonnet and Newcomb (1965) in the root cells of *Raphanus sativus* (radish) and later by Favali and Gerola (1968) in the phloem parenchyma of foliar veins in *Brassica chinensis*. Jorgensen *et al.* (1977) detected them in foliar veins in *Capparis cynophallophora*. Iverson and Flood (1970) found that such structures are common occurrence in the Cruciferae (Brassicaceae). According to Behnke (1977) these organelles are typical of Brassicaceae and Capparaceae.

Behnke (1972, 1977, 1981) in a series of observations since about three decades ago, has demonstrated ultrastructural details of sieve element plastids, which can be used in plant taxonomy. Behnke studied more than 2500 species from 380 families with regard to this character. There are broadly two types of such plastid elements, one accumulating starch (S-type) and the other accumulating protein (P-type). There are, however, certain groups of plants-Rafflesiaceae, Crassulaceae and some species of Ulmaceae, Moraceae and Urticaceae in which the plastids have neither starch nor protein accumulations, and they are called S_0-type. S- type sieve element plastids with starch accumulation in grains of different sized are the commonest type in flowering plants. About 65 per cent of these have such plastids in the sieve tube elements (Behnke, 1977). The subclasses Ranunculidae, Hamamelididae and Dilleniidae have without exception S-type sieve element plastids, while in the Rosidae a great majority of the orders have this type. Half the members of the Magnoliidae have S-type and half P-type elements. In Caryophyllidae majority of the members show P- type sieve element plastids (Figure 9.3).

P-type plastids are further differentiated on the number and shape of the crystalloids and the nature of the filaments surrounding them.

1. P I subtype: The plastids contain single crystalloids of different sizes and shapes and/or irregularly arranged filaments. PI subtype sieve element plastids are present in Magnoliales, Laurales and Aristolochiales.
2. P II subtype The plastids contain several cuneate crystalloids oriented towards the centre of the plastid. PII subtype sieve element plastids are present in Monocotyledons.
3. PIII subtype: The plastids contain a ring-shaped bundle of filaments. These plastids are seen only in Centrospermae.
4. P IV subtype: The plastids contain a few polygonal crystalloids of various sizes. This type of plastids are seen only in Fabales.
5. PV subtype: The plastids contain many crystalloids of different sizes and shapes. It is seen in the order Ericales and family Rhizophoraceae.
6. PVI subtype: The plastid contains a single circular crystalloid. It is found in family Buxaceae.

Behnke (1974) investigated the order Rhamnales and found that the family Rhamanaceae had S-type while Vitaceae and Leeaceae had P-type sieve element plastids. The heterogeneity of the Magnoliidae as to its plastid types is attributed to its basal position in the evolution of flowering plants (Behnke, 1975). P-type are considered to be ancestral and S-type has been thought to be derived by loss of protein. P I subtype is held to be the most primitive among P-subtypes of flowering plants.

The best use of plastids has been made in elucidating the interrelationships and circumscribing the order Caryophyllales (Centrospermae) (Behnke, 1976). All the families kept in this order show P III subtype sieve element plastids. These families are : Aizoaceae, Nyctaginaceae, Phytolaccaceae, Portulacaceae, Caryophyllaceae and Molluginaceae. In earlier classification other families like Polygonaceae, Vivianiaceae,

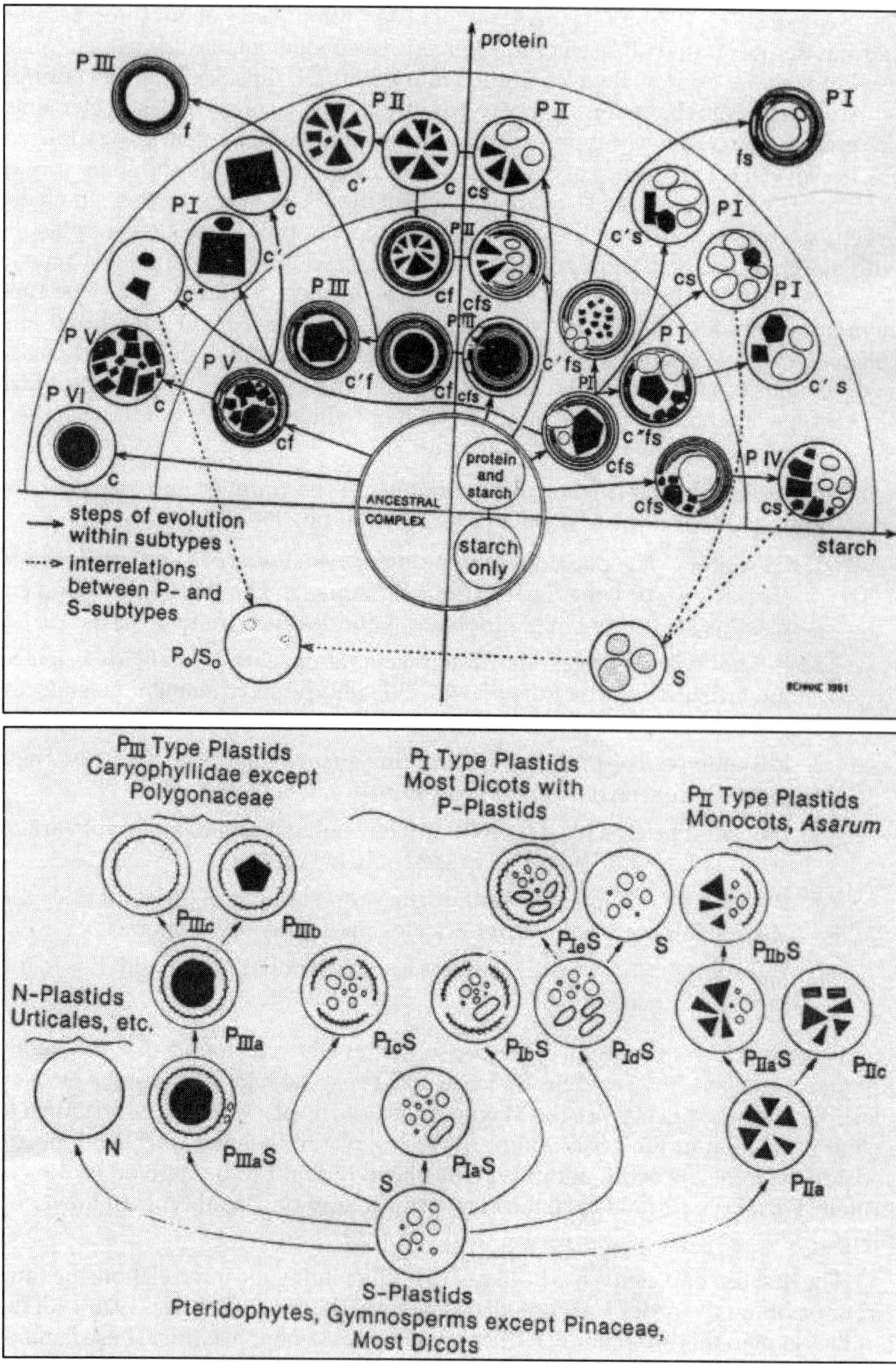

Figure 9.3: Sieve Element Plastids in Angiosperms

Bataceae, Gyrostemonaceae, Plumbaginaceae, Fouquieriaceae, Frankeniaceae, Rhabdodendraceae and Theligoniaceae have been kept in Centrospermae but in recent classifications these have been removed from Centrospermae because they show S-type sieve element plastids.

Embryology in Relation to Taxonomy

The role of embryology in taxonomy was first brought to prominence by the German embryologist Schnarf in 1931. Subsequently, enormous work has been done on this aspect and the data collected has proved to be of considerable taxonomic significance. Often a single embryological character can mark out the family, *e.g.*, Pseudoembryo sac in the Podostemaceae, composite endosperm in the Loranthaceae and formation of a single pollen grain from a pollen mother cell in the Cyperaceae. The importance of embryology in taxonomy has been summarized by many embryologists like Maheshwari (1950), Johri (1963), Davis (1966), Poddubnaja-Arnoldi (1976), Herr (1984), Batygina and Yakovlev (1980-1990) and Johri *et al.* (1992).

Embryological Features of Taxonomic Importance

1. *Anther*: Number of vertical rows of cells of archesporium, type of wall development (Basic, Monocotyledonous, Dicotyledonous and Reduced), the tapetum glandular or amoeboid, number and ploidy of nuclei in tapetal cells, successive or simultaneous cytokinesis, types of tetrad.
2. *Pollen grain*: Number and position of germ pores, number of cells at the shedding stage.
3. *Ovule*: Type of ovule (orthotropous, anatropous, campylotropous, amphitropous or circinotropous), number of integuments, micropyle formed by one (inner or outer) or two integuments, nucellus crassinucellate, pseudocrassinucellate or tenuinucellate, presence or absence of endothelium, epistase, hypostase, aril, caruncle, obturator etc.
4. *Archesporium*: Number or archesporial cells, extent of parietal tissue.
5. *Megasporogenesis and development of embryo sac*: Shape of the megaspore tetrad, position of functional megaspore in the tetrad; embryo sac development mono, bi- or tetrasporic, number of nuclear divisions.
6. *Mature embryo sac*: Number and organization of cells in the embryo sac.
7. *Fertilisation*.
8. *Endosperm*: Nuclear, Cellular or Helobial, presence or absence of haustoria, seeds albuminous or exalbuminous.
9. *Embryo*: Development type (Asterad, Onagrad, Chenopodiad, Solanad and Caryophyllad), presence or absence of suspensor haustoria.
10. *Seed coat*: Number of layers and their derivation (from one or two integuments).
11. *Special features*: Apomixis, polyembryony.

Application of Embryology to Problems of Taxonomy

Embryological characters are useful in delimitation of plant groups at all levels. At the highest level Angiosperms are characterized by double fertilization (syngamy and triple fusion) which is not seen in any other division. Similarly Angiosperms differ from Gymnosperms in that the nutritive tissue for development of embryo in Angiosperms is a post fertilization phenomenon while in Gymnosperms it is a pre-fertilization character. The Dicotyledons and Monocotyledons differ in the cytokinesis of pollen mother cells after meiosis. The Dicotyledons are characterized by the simultaneous cytokinesis while Monocotyledons are characterized by the successive cytokinesis.

Within Dicotyledons and Monocotyledons the subclasses are also differentiated on embryological grounds. The Asteridae are characterized by unitegmic and tenuinucellate ovules while the Caryophyllidae are characterized by bitegmic and crassinucellate ovules.

Embryological data supplement other evidence in many cases. Polygonaceae has been associated by various taxonmists with the Centrospermae group. Polygonaceae differs embryologically from other Centrospermae group. Polygonaceae are characterized by the orthotropous ovules while Centrospermae have either Campylotropous or amphitropous ovules.

Some Families Marked Out by their Embryological Features

Podostemaceae

An exclusively tropical family, Podostemaceae includes perennial aquatic herbs, which grow attached to stones in running water. The family shows interesting embryological features. A unique feature is the formation of pseudoembryo sac due to the disintegration of nucellar cells below the embryo sac (Figure 9.4B). A combination of the following characters further mark out this family.

1. Occurrence of pollen grains in pairs (Figure 9.4A),
2. Anatropous, bitegmic and tenuinucellate ovule with micropyle formed by the outer integument (Figure 9.4B),
3. Bisporic embryo sac,
4. Solanad type of embryogeny,
5. Prominent suspensor haustoria (Figure 9.4C),
6. Absence of antipodal cells (except in *Dicraea*),
7. Absence of triple fusion and consequently endosperm,
8. Presence of pseudoembryo sac.

Onagraceae

The family Onagraceae is characterized by the presence of Oenothera type of embryo sac development. Oenothera type of embryo sac (Figure 9.5) development is seen in all the members of the family Onagraceae and is not found in any other family. Further of the four megaspores formed after meiosis, the chalazal three

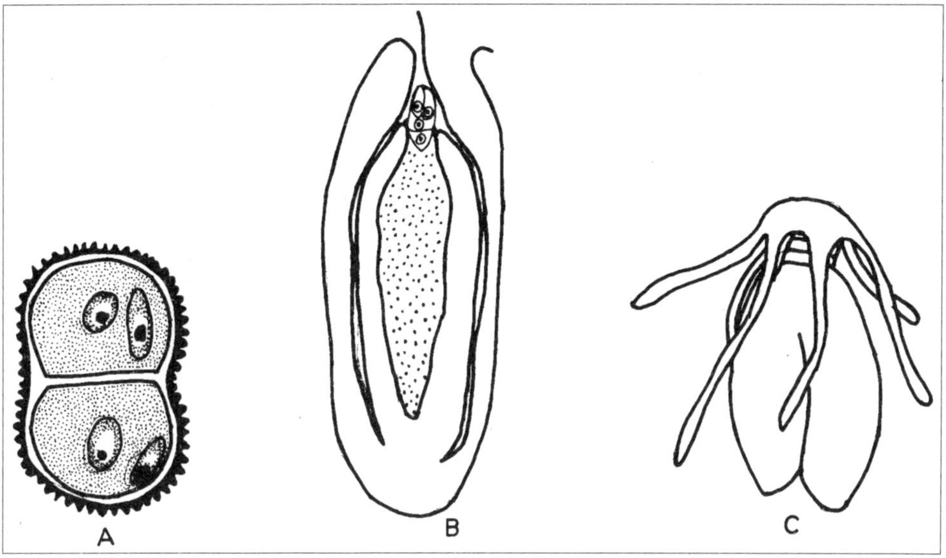

Figure 9.4

megaspores degenerate, while the micropylar megaspore develops into embryo sac. The family is also characterised by the absence of antipodals. Onagraceae are further characterized by a diploid endosperm.

Cyperaceae

The development of microspore from the pollen mother cell in the family Cyperaceae is quite unique. In all other Angiosperms the pollen mother cell divides by meiosis, resulting in four haploid nuclei, each of which forms a functional microspore. The members of the Cyperaceae exhibit a special mode of pollen development. Of the four nuclei formed after meiosis, only one function. The functional nucleus remains in the centre and the three non-functional nuclei are cut off on one side of the cell, and they do not form pollen grains. The functional nucleus divides to form a vegetative cell and a generative cell. The non-functional nuclei eventually degenerate (Figure 9.6).

Loranthaceae

Members of the Loranthaceae show some embryological features, which are not seen in any other Angiosperms. In this family members do not show any differentiation in ovules and placenta. Here no ovules are present as such and the placental ovular complex is known as a mamelon. The integuments have disappeared and there is a gradual reduction in the lobing and differentiation of mamelon. Finally in some genera even the mamelon is absent and the archesporial cells differentiate at the base of the ovarian cavity. The embryo sacs ascend to various heights in the style. The family is also characterized by the composite endosperm formed by several embryo sacs.

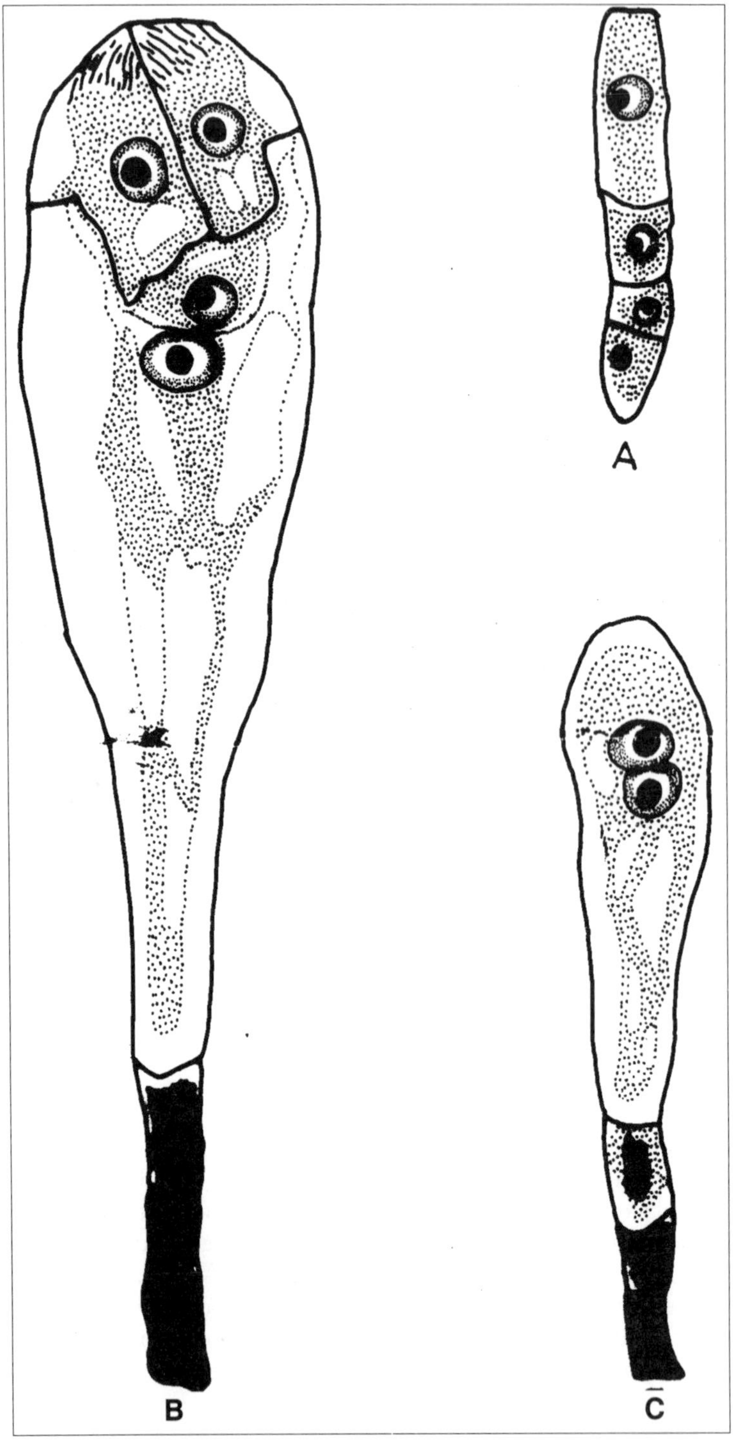

Figure 9.5: Embyro Sac Development in Onagraceae

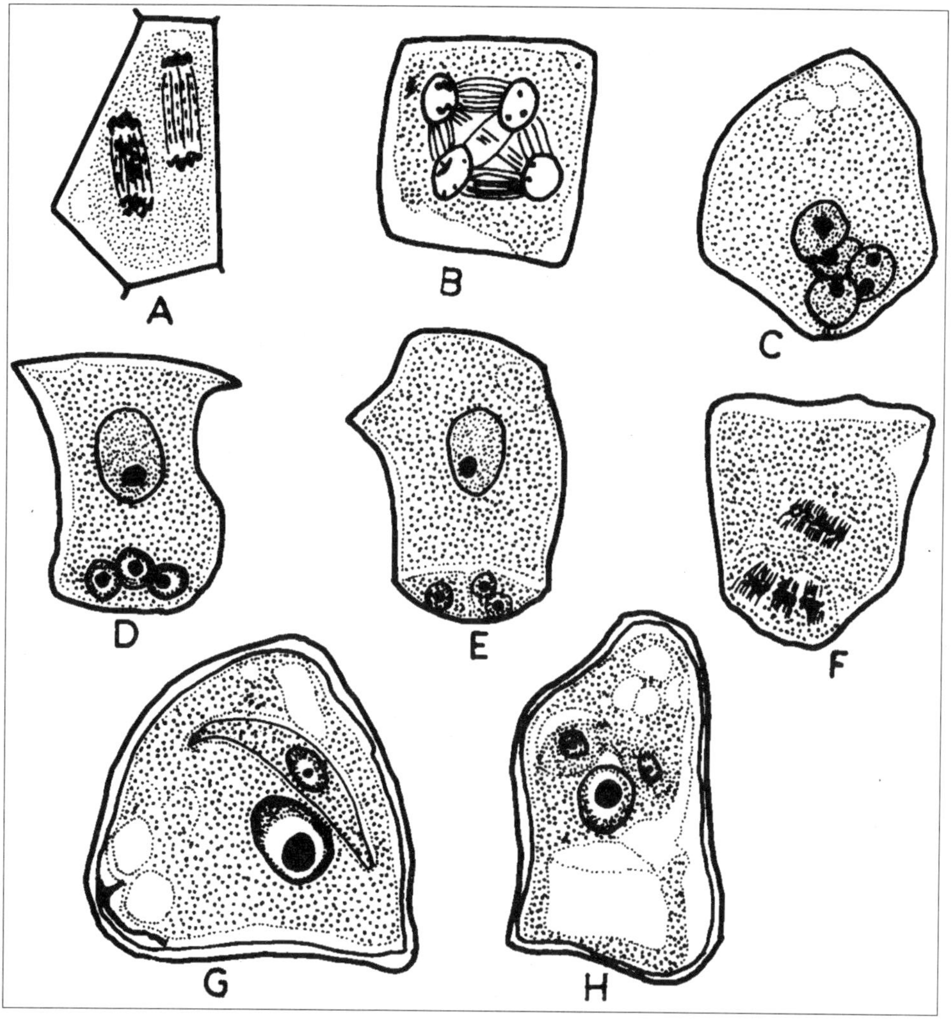

Figure 9.6: Pollen Development in Cyperaceae

Some other examples of the value of embryology in Taxonomy

Paeonia

In most of the classical systems of classification the genus *Paeonia* is kept in the monogeneric tribe *Paeonieae* of the Ranunculaceae. The vascular anatomy and certain characters of leaves and fruits led Worsdell (1908) to conclude that *Paeonia* constitutes a natural and independent group and should be assigned to a separate family Paeoniaceae which is closer to Magnoliaceae than Ranunculaceae. Kumazawa (1935) studied its wood anatomy and confirmed this conclusion. In the structure of exine Wodehouse (1936) found *Paeonia* to differ from the rest of the Ranunculaceae. Gregory (1941) noticed that the basic chromosome number of *Paeonia* is five in contrast to seven, eight or nine in the Ranunculaceae. Eames (1951) also pointed out that the floral anatomy of *Paeonia* is quite different.

Embryological data also support the view of removing *Paeonia* from the Ranunculaceae to a separate family Paeoniaceae. The most peculiar embryological feature of this genus is its embryogeny (Yakovlev and Yoffe, 1957; Murgai, 1959, 1962). According to Yakovlev and Yoffe (1957) the zygote undergoes repeated nuclear

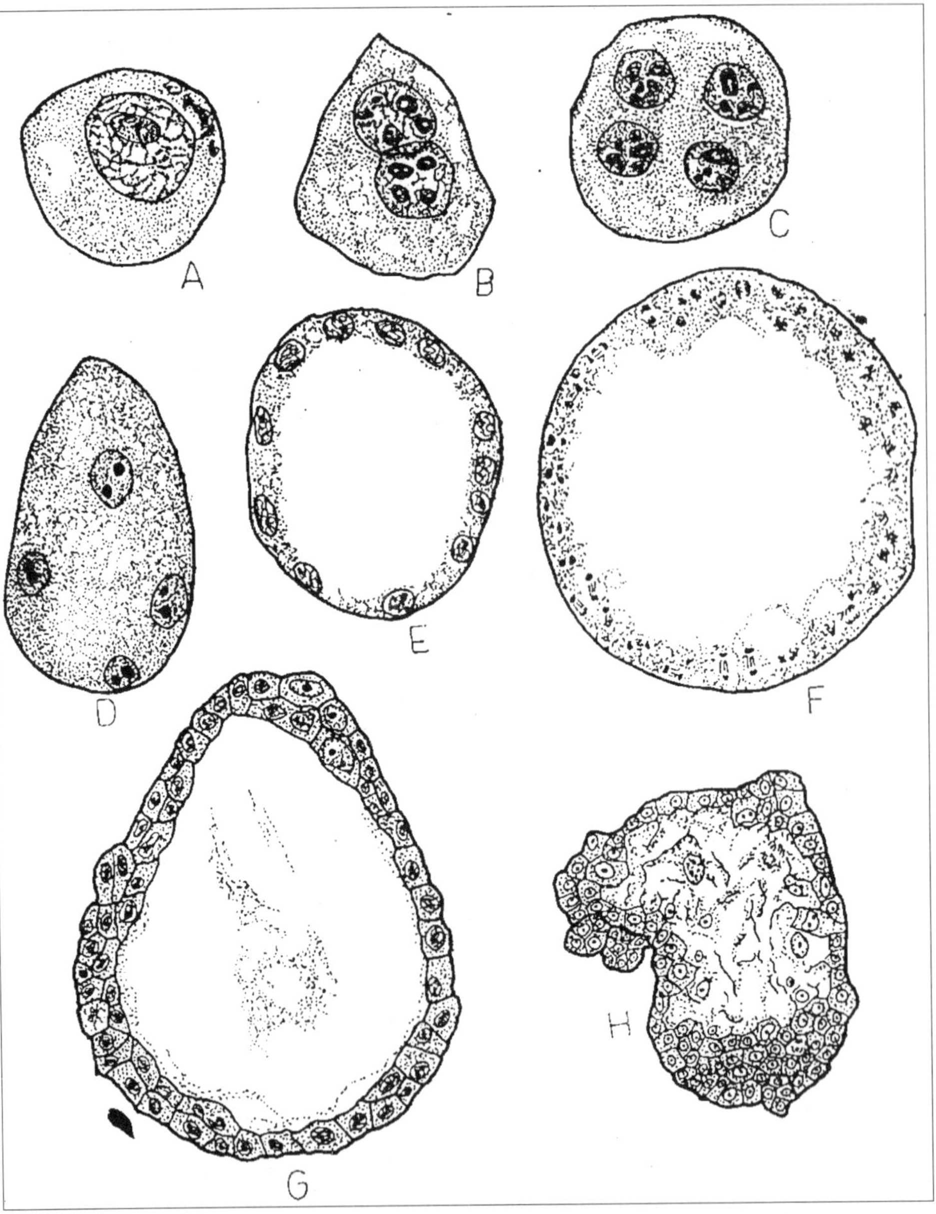

Figure 9.7: Embryo Development in Paeoniaceae

divisions forming coenocytic structure (Figure 9.7 A-C). Later, the nuclei become lodged in the peripheral layer of cytoplasm around a large central vacuole (Figure 9.7D). Now, wall formation occurs, and the coenocyte becomes cellular, except in the central part (Figure 9.7F) of which one matures into adult embryo.

Other embryological features which justify the separation of *Paeonia* from Ranunculaceae to a separate family are given in Table 9.1.

Table 9.1: Embryological Differences Between *Paeonia* and Ranunculaceae

Feature	*Paeonia*	*Ranunculaceae*
Stamens	Spirally arranged, centrifugal	Spirally arranged, centripetal
Anther	Multilayered endothecium, mostly 2-layered tapetum	One-layered endothecium and one-layered tapetum
Pollen	Reticulately pitted exine, large and elongate generative cell	Granular, papillate or smooth exine, small lenticular generative cell.
Female archesporium	Multicelled, many megaspore mother cells function	Uni- or multicelled, one cell functions
Antipodal cells	Persistent, not polyploid	Persistent (ephemeral in *Adonis*), nuclei one or more than one, polyploid.
Embryogeny	Unique, Free nuclear	Onagrad or rarely Solanad type
Seed	Arillate	Non arillate

Trapa

Bentham and Hooker (1883) placed the genus *Trapa* in the family Onagraceae, while Raimann (1898), Engler and Diels (1936) and Hutchinson (1959) included it in the unigeneric family Hydrocaryaceae now called Trapaceae. Eames (1953) remarked that on the basis of anatomical evidence *Trapa* does not belong to Onagraceae. The embryological data presented in Table 9.2 also support the view that *Trapa* should be removed from Onagraceae to Trapaceae.

Table 9.2: Embryological Differences Between *Trapa* and Onagraceae

Feature	*Trapa*	*Onagraceae*
Pollen grain	Pyramidal with 3 much folded meridional crests	Bluntly triangular and basin shaped
Ovary	Semi-inferior, bilocular with a single pendulous ovule in each chamber	Inferior, mostly trilocular with many ovules per chamber on an axile placenta
Embryo sac	Polygonum type	Oenothera type
Endosperm	Absent	Present, nuclear
Embryo	Solanad type	Onagrad type
Suspensor	Well developed suspensor haustorium	Short and inconspicuous
Cotyledons	One cotyledon extremely reduced	Cotyledons equal
Fruit	Large, one-seeded drupe with prominent spines	Loculicidal capsule

Basellaceae

Several divergent views have been expressed about the systematic position of Basellaceae, a small family of herbaceous twining vines that occur chiefly in tropical America and Asia. Eichler (1878), Volkens (1893) and Lawrence (1951) include the family in Centrospermae close to Portulacaceae. A similar assignment is offered by Bessey (1915), Gunderson (1950), Takhtajan (1997) and Cronquist (1989), although these systematists have adopted the alternative ordinal name, Caryophyllales. Hutchinson (1973) includes the family among group of advanced Caryophyllaceous families which he elevates to a separate order, Chenopodiales. Hooker (1880) includes it in Chenopodiaceae as the subfamily Baselleae and Franz (1908) lowers it to a tribal rank, placing it in subfamily Montioideae of Portulacaceae.

Maheswari Devi and Pullaiah (1975) investigated the life history of *Basella rubra* and summarised the embryological features of the family Basellaceae. They also offer a comparison of those families included in Centrospermae in Table 9.3.

Basellaceae and Chenopodiaceae differ markedly embryologically, and the inclusion of Basellaceae as a subfamily of Chenopodiaceae is untenable. This conclusion is also supported by anatomical features summarized by Metcalfe and Chalk (1950). For example, stems in Basellaceae contain a single ring of vascular strands with intraxylary phloem in the larger bundles. Chenopodiaceae, on the other hand, are characterized by the development of polycyclic, anomalous secondary vascular bundles.

Basellaceae and Portulacaceae share many embryological features (Table 9.3) supporting a close relationship, as also suggested by Takhtajan (1997). However, to include Basellaceae as a tribe in Portulacaceae does not appear justified in view of certain differences.

Since the Basellaceae show all the embryological features which predominate in Caryophyllales, the retention of Basellaceae as one of the independent families in Caryophyllales seems to be fully justified on embryological grounds.

Gentianaceae and Menyanthaceae

Benthem and Hooker (1873-1876) and Rendle (1952) split the Gentianaceae into two subfamilies: (a) Menyanthoideae including the marshy plants, and (b) Gentianoideae including all the others. Wettstein (1935) raised Menyanthoideae to a family rank, which was accepted by Don (1938) and Lindsey (1936). This was supported by both vegetative (Perrot, 1897) and floral anatomy (Lindsay, 1938) and embryology (Soueges, 1943; Maheswari Devi, 1962).

Morphological and embryological dissimilarities of the subfamilies Menyanthoideae and Gentianoideae are tabulated in Table 9.4.

Maheswari Devi (1962) is of the opinion that these differences fully justify a family rank for the Menyanthoideae which would be better designated as Menyanthaceae. The name Gentianaceae should be retained for Gentianoideae.

Table 9.3: Embryology of Phytolaccaceae, Aizoaceae, Portulacaceae, Basellaceae, Caryophyllaceae and Chenopodiaceae (Adapted after Maheswari Devi and Pullaiah, 1975)

	Phytolaccaceae	*Aizoaceae*	*Portulacaceae*	*Basellaceae*	*Caryophyllaceae*	*Chenopodiaceae*
Anther	Tetrasporangiate	Tetrasporangiate	Tetrasporangiate	Tetrasporangiate	Bisporangiate	Tetrasporangiate
Anther wall development	–	Basic type	Monocotyledonous	Monocotyledonous	Mono- or dicotyledo-nous	Monocotyledonous
Endothecium	Fibrous	Fibrous	Fibrous	Fibrous, multiseriate	Fibrous	Fibrous
Tapetum	Glandular, cells bi- or multinucleate	Glandular, cells bi-four-nucleate	Glandular, cells bi- or multinucleate	Glandular, cells multinucleate	Glandular, cells binucleate	Glandular, cells uni- or binucleate; occasionally amoeboid
Tetrads	Tetrahedral, isobilateral	Tetrahedral, isobilateral, decussate	Tetrahedral, isobilateral	Tetrahedral, isobilateral	Tetrahedral, decussate	Tetrahedral, isobilateral decussate
Pollen grain	Two- (*Phytolacca*) or three-celled (*Rivina*)	3-celled	3-celled	3-celled	3-celled	3-celled
Ovule	Anacampylotropous, bitegmic crassinucellate	Anacampylotropous, bitegmic crassi-nucellate, funicular obturator develops	Anacampylotropous, bitegmic crassinucellate	Ana- or campylo-tropous, bitegmic crassinucellate	Campylotropous, bitegmic crassi-nucellate, nucellar beak conspicuous	Campylotropous, bitegmic crassi-nucellate nucellar beak conspicuous
Microphyle	Formed by inner integument	Inner integument	Inner integument	Inner integument	Inner integument	Inner integument
Air-space between the integuments at their base	Present	Present	Present	Present	Present	Present
Archesporium	2- or 3-celled	Multicelled	2- or 3-celled	2- or 3-celled	Uni- or multicelled	2- or 3-celled

Contd...

Table 9.3–*Contd...*

	Phytolaccaceae	*Aizoaceae*	*Portulacaceae*	*Basellaceae*	*Caryophyllaceae*	*Chenopodiaceae*
Embyro sac	Polygonum type	Polygonum, Endymion, Drusa, Adoxa or Penaea	Polygonum	Polygonum	Polygonum	Polygonum, *Allium* type in *Suaeda maritime*
Antipodal cells	Persistent, secondary multiplication in *Rivina*	Ephemeral	Ephemeral	Ephemeral	Ephemeral supernumerary due to divisions	Ephemeral
Endosperm	Nuclear type	Nuclear, chalazal, haustorium in *Hereoa hesparantha*	Nuclear	Nuclear	Nuclear, in some taxa superficial extension develops	Nuclear
Embryogeny	Caryophyllad type, massive suspensor	Solanad, occasional suspensor polyembryony	Onagrad or Caryophyllad, occasional polyembryony	Chenopodiad	Caryophyllad, basal cell of suspensor haustorial	Chenopodiad, suspensor massive
Seed	Small, exotestal, perispermous	Exotestal, and endotegmic, perispermous	Small, exotestal, albuminous	Small, exotestal, perispermous	Small, exotestal, perispermous	Small, exotestal and endotegmic, perispermous

Table 9.4: Differences Between Menyanthoideae and Gentianoideae

Feature	*Menyanthoideae*	*Gentianoideae*
Habit	Aquatic or marshy	Tetrrestrial
Morphology	Leaves alternate, corolla valvate or imbricate	Leaves opposite and decussate, corolla twisted
Anatomy	Vascular bundles collateral, sieve tubes absent in phloem	Vascular bundles bicollateral or amphi-cribal, sieve tubes present in phloem
Anther tapetum	Secretory type, cells bi- or multi-nucleate, nuclear fusion common	Secretory (*Exacum*) or amoeboid (*Canscora*), cells remain uninucleate
Placenta	Suppressed parietal placenta	Well developed parietal placenta, some times the placenta fuse at the centre to form a bilocular ovary
Endothelium	Present	Absent
Endosperm	Cellular	Nuclear
Embryogeny	Asterad type	Solanad type

Palynology and Taxonomy

Pollen morphology offers reliable data and evidence of taxonomic relationships. Palynological evidence may be used to place taxa of uncertain affinities, to suggest rearrangements, withdrawals and separations as well as corraborating other lines of evidence. Sometimes, the study of fine structural details in fossil pollen opens up possibility of new sources of phylogenetic evidence.

The major characters of pollen grains that are used in solving taxonomic and phylogenetic problems are (1) Apertural patterns, (2) Exine stratification and ornamentation, (3) Pollen association and (4) Pollen nuclear number.

The taxonomic and evolutionary importance of pollen morphology may be at specific, generic or higher levels. In many cases the type of pollen of a taxon is characteristic and constant; such a taxon is called *Stenopalynous* or unipalynous (Nair, 1970) and may be exclusive of that group, *e.g.*, thick walled pollen grains of the Gyrostemonacee of Australia. In other cases, the types of pollen may vary considerably in size, aperture, stratification of exine etc. Such taxa are termed *eurypalynous* or multipalynous. Stenopalynous taxa are generally considered to be very natural. Ascelpiadaceae, Asteraceae, Poaceae, Brassicaceae and Lamiaceae are some of the stenopalynous families.

Eurypalynous condition, seen in Malvaceae, is primitive while stenopalynous condition, seen in Poaceae, is an advanced character.

Apertural Patterns

By using the NPC system, formulae can be derived to individual taxa and families. Families with more closely similar formulae must be considered more closely related than those with less similar formulae.

Engler and Diels (1936) included Canellaceae in the order Parietales. When NPC formulae of all the families included in the parietales are observed, they appear to be more or less uniform having $N_3P_4C_5$ (345). But Canellaceae appears to be an exception - with the formula 133. Subsequently Melchoir (1964) divided partietales into two orders the Violales in which the NPC is essentially 345 throughout and others in Guttiferales.

The genus *Theobroma* provides an example of a genus with two distinct types of pollen. One is small, flattened and is found in the majority of the species including *T. cacao.* The other type is larger, more elongate. This is one of the reasons for the recent removal of the species with larger pollen grains to a distinct genus *Herrania.*

Caltha palustris var. *palustris* and *C. palustris* var. *alba* can readily be distinguished by their aperture.

C. palustris var. *palustris*	*C. palustris* var. *alba*
Tricolpate	Pantoporate
Position of apertures equatorial	Position of apertures global
Size of apertures - large	Size of apertures - small

In the genus *Sesamum,* the number of colpi is species specific. Eleven in *S. indicum,* nine in *S. prostratum* and eight in *S. laciniatum.* Similarly *Butomopsis umbellatus* can be differentiated from *B. lanceolata* by the nature of apertures, the former having monocolpate and the later pantoporate grains.

Salix and *Populus* of Salicaceae can readily be distinguished from one another on the basis of aperture. The former has 3-colpate pollen and the latter inaperturate one.

Similarly *Phytolacca* and *Rivinia* of Phytolaccaceae can be distinguished from one another on pollen apertural patterns. *Phytolacca* has 3-zonocolpate pollen while *Rivinia* has 5-colpate pollen.

The genus *Nelumbo* has till recently been included in Nymphaeaceae. *Nelumbo* has 3-colpate pollen while Nymphaeaceae have uniformly monosulcate pollen. Hence *Nelumbo* has been segregated into a separate family Nelumbonaceae.

Bentham and Hooker (1862-83) and other earlier taxonomists have included *Alisma, Butomus* and *Butomopsis* in the family Alismataceae. But Hutchinson (1969, 1973), Cronquist (1988) and Takhtajan (1980) have segregated *Alisma* into Alismataceae and *Butomus* and *Butomposis* into Butomaceae. This is supported on Palynological grounds. *Alisma* have pantoporate pollen while *Butomus* and *Butomopsis* have monocolpate one.

Apertural characters are also useful in understanding the phylogeny. The monosulcate (uniaperturate) pollen found in some Ranales indicate primitiveness, Triaperturate and polyaperturate pollen are found in advanced families like Asteraceae.

Exine Sculpturing

The exposed surface-details of the pollen wall constitute the sculpturing. Some of the more important types are: Psilate (smooth), foveolate (pitted), fossulate (grooved),

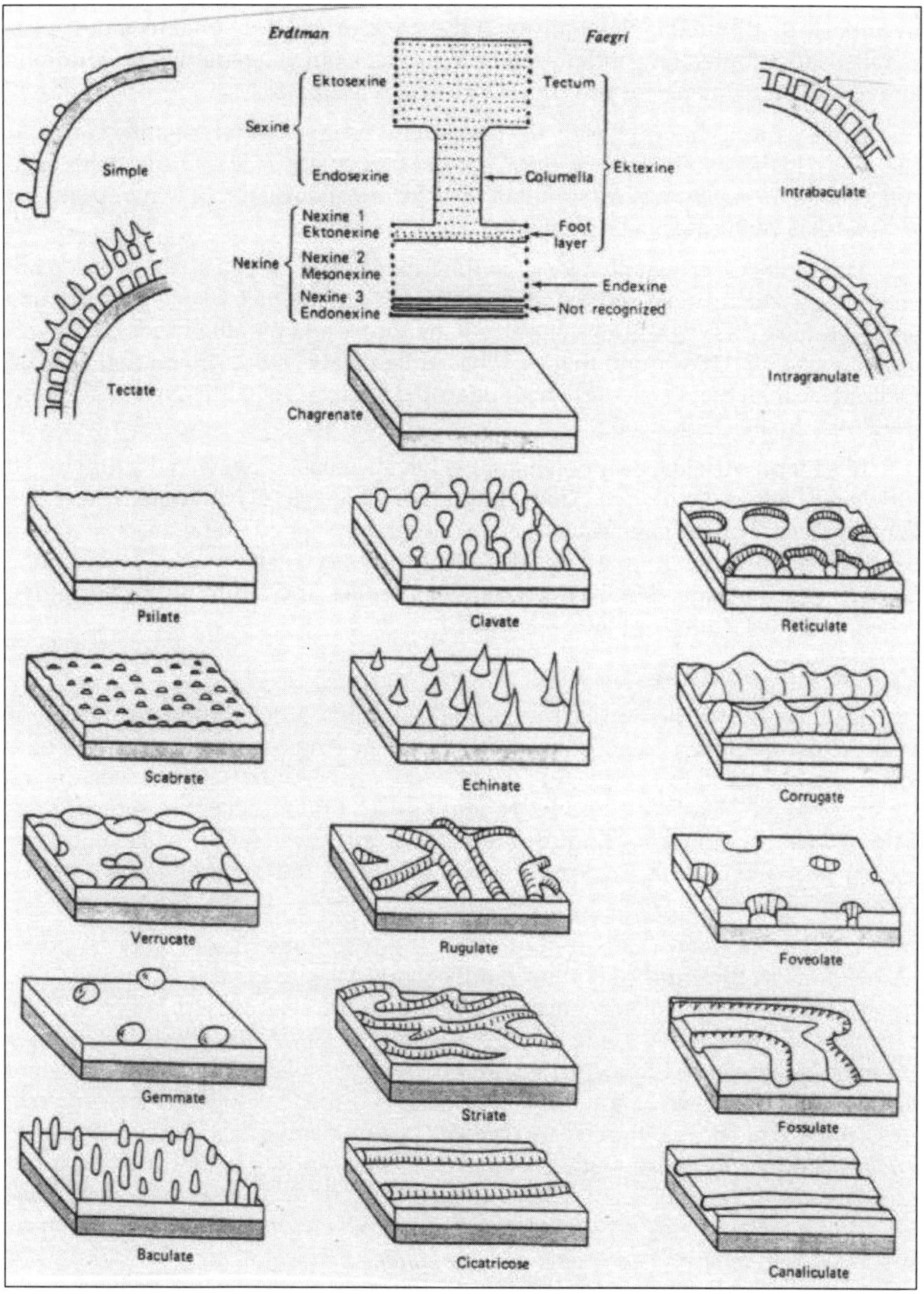

Figure 9.8: Exine Patterns in Angiosperm

scabrate (very fine projections), verrucate (warty), baculate (rod-like elements with swollen tips), gemmate (sessile pillar), echinate (spiny), rugate (elongate elements irregularly distributed tangentially over the surface), striate (elongate, more or less parallelly distributed tangentially over the surface), punctate (minute perforations) and reticulate (elements forming open network) (Figure 9.8).

Exine patterns have occasionally been found useful in differentiating species of the genus. In the genus *Bauhinia, B. racemosa* has reticulate, *B. purpurea* has reticulate-tuberculate, *B. malabarica* has spinulate *B. acuminata* has pilate, *B. krugii* striate and *B. retusa* has verrucate pollen grains.

The genus *Cicer,* usually placed in the tribe Vicieae of the Fabaceae, has many characters, which suggested that this position is anomalous. Clarke and Kupicha (1976) studied the pollen grains of *Cicer,* of the Viceae and the allied tribes, Trifolieae and Ononideae. They found that *Cicer* has pollen that is more similar to that of the Ononideae than that of Viceae. They suggested the transfer of the genus *Cicer* from the Viceae to the Ononideae.

In general, Angiosperm pollen grains have a massive exine and a thin intine. However certain taxa (*e.g.,* Amaryllidaceae, Cannaceae, Iridaceae, Musaceae, Zingiberaceae, Costaceae, Marantaceae, Strelitziaceae and Heliconiaceae among Monocotyledons) are unique in having a highly reduced exine and elaborate intine. In *Heliconia,* the exine is relegated to a few spinules and a thin layer over intine, which is almost a hundred times as thick.

Pollen Association

In majority of the Angiosperms pollen grains occur singly. However, in Podostemaceae pollen grains occur in pairs. In some Angiospermous families pollen grains are clustered in Tetrads. In Dicotyledons such associations are seen in such diverse families like Winteraceae, Monimiaceae, Annonaceae, Droseraceae and Mimosaceae. In Monocots, they are restricted to advanced families like Juncaceae. Extreme cohesion of pollen is met with Asclepiadaceae and Orchidaceae, where they form pollinia.

Vij and Kashyap (1978) pointed out the taxonomic importance of pollen association in the multipalynous family Orchidaceae. Of the 50 species they investigated, single pollen grains were found in the relatively primitive groups (*Paphiopedilium villosum, Cyprepedium cordigerum, Cephalanthera ensifolia*), tetrads in the tribe Neottieae and Epidendrieae and perfect massulae in Orchideae and some Neottieae. Poddubnaja Arnoldi (1976) has also reported that pollen grains are single in *Cypripedium insigne,* tetrads in *Calanthe veitochii* and massulae in *Platanthera bifoliata.* Pollen grains are single in diandrous forms, and some monandrous forms (*Cephalanthera, Vanilla*). They remain attached in tetrads as in *Spiranthes australis, Aa achalesis, Bromhedia finlaysoniana, Calanthe* and *Polystachya flavescens* or form massulae as in *Epigonium roseum, Liparis viridiflora, Platanthera* and *Peristylus spiralis* or pollinia as in Monandrae (see Johri *et al.,* 1992).

Similarly the family Mimosaceae pollen grains are shed singly, in tetrads or pollinia. In *Neptunia, Leucaena, Prosopis* and *Desmanthus* pollen grains are shed singly.

In *Mimosa pudica* pollen grains are shed in tetrads but in *Mimosa hamata* pollen grains are shed in single tetrads as well as twin tetrads. They are shed together in masses of eight or sixteen pollen grains in *Acacia dealbata, A. farnesiana* and *A. leucophloea.* All the microspores (in a sporangium) together form a single pollinium in *Albizzia lebbeck, Acacia baileyana* and *Calliandra.*

Pollen Nuclear Number

In Angiosperms pollen grains are shed either at 2-nucleate stage or 3-nucleate stage. This character is valuable in taxonomy and phylogeny of Angiosperms.

Brewbaker (1967) made and extensive survey of about 2000 species belonging to 265 families. According to Brewbaker two-nucleate pollen grains are more primitive in Angiosperms as a whole and within individual taxa, three-nucleate pollen grains have arisen from the two-nucleate ones.

In the Monocotyledons, the three-nucleate pollen grains occur in highly reduced wind pollinated groups, aquatics and a chlorophyllous saprophytes. In the Dicotyledons, however, there is no definite pattern. Certain orders, *e.g.*, Caryophyllales, Plumbaginales, Polygonales and Asterales, have consistently tri-nucleate pollen grains (Dahlgren, 1975). Magnoliales, Laurales and Violales have consistently two-nucleate pollen grains. Most other orders are quite heterogeneous in having both two and three-nucleate ones. Both these types may occur in one and the same genus, as has been observed in *Burmannia, Lobelia* and *Ipomoea* (Brewbaker, 1967).

The important recently discovered connection between bi- and tri-nucleate pollen grains and types of self-incompatibility systems (see Brewbaker, 1957) opens up further possibilities. Particularly in the dicotyledonous taxa the groups with binucleate pollen grains tend to have the gametophytic type of self-incompatibility and those with trinucleate pollen grains the sporophytic type. In Lamiaceae Kooiman (1972) has also demonstrated the correlation between binucleate and tricolpate pollen grains and between trinucleate and hexacolpate pollen grains.

Cytology in Relation to Taxonomy

Cytotaxonomy utilizes cytological characters in solving taxonomical problems. Solbrig (1968) indicated that most Cytotaxonomic studies have largely made use of only two nuclear aspects: (1) number, shape and size of chromosomes and (2), the behaviour of chromosomes during mitosis and meiosis. Ehrendorfer (1964) has reviewed the previous work pertaining to the taxonomic role of chromosome studies.

Chromosome Number

There is a great variation in the number of chromosomes in vascular plants. All individuals within a species have the same chromosome number. Somatic cells of *Haplopappus gracilis, Brachyscome lineariloba* (Asteraceae) and *Colpodium versicolor* (Poaceae) show 4 chromosmes per cell, *Arabidopsis thaliana* has 2n = 10 chromosomes while *Poa littrosa* (Poaceae) show 530 chromosomes, *Kalanchoe* and *Echeveria* (both Crassulaceae) species show 500-520 chromosomes, *Morus nigra* (Moraceae) 308 chromosomes and *Ophioglossum reticulatum* have 1260 chromosomes in their somatic cells.

The great diversity in chromosome number and their constancy in species provide an important character for taxonomic groupings of large number of plants.

In higher plants there are three main different cytological types-diploid, polyploid and aneuploid. Certain groups of vascular plants possess a constant number of chromosomes. Genera such as *Pinus* and *Quercus* have n = 12 chromosomes in all the known species and can be delimited from other related genera on the basis of chromosome number.

When chromosome numbers in various members of a taxon are in the proportion of exact multiples, the series is described as euploidy. Several such examples can be found in the flowering plants and the ferns. In Malvaceae for instance, the somatic numbers in various species range from 10, 15, 20, 25....40, from 12, 18, 24 to 30; from 14, 28, 42, 56 to 84 and so on. In the genus *Taraxacum* (Asteraceae), there are species with 2n=16, 24, 32, 40 and 48 chromosomes. Various members of such an euploid series may be unified by a basic number of x which is the gametic number of a diploid species and other species can be described triploids (3x), tetraploids (4x), hexaploids (6x)....Polyploids (nx). Thus one of the series in Malvaceae has x=5 while the others have x = 6, 7 etc. in *Taraxacum* the basic number is x = 8.

In the living species of *Chlorophytum* (Liliaceae), for example, the chromosome numbers vary from 14 to 28, 42, 56, 84 etc. and also from 16 to 32. This means that *Chlorophytum* has two basic numbers x = 7 and 8. However, from the detailed analysis of the meiotic behaviour of chromosomes in one of the species *C. laxum* Naik (1976a) has shown that the two base numbers 7 and 8 can be considered as secondary, most probably derived from primary basic number x = 4.

In certain groups, the closely related species are cytologically distinct, one being a diploid and the other a tetraploid. Such clearly related pairs are termed polyploid pairs. These have been shown experimentally in *Nasturtium officinale* (2n=32) and *N. macrophyllum* (2n = 64); *Cardamine hirsuta* (2n = 26) and *C. flexuosa* (2n =32) of the Cruciferae and *Saxifraga hyperborea* (2n = 26) and *S. rivularia* (2n =52) of the Saxifragaceae.

Hybridization plays a major role in evolution. In Brassicaceae, *Brassica oleracea* (2n=18) and *B. rapa* (2n = 20) have been shown to hybridize in nature to give rise to *B. napus* (2n = 38) after polyploidy. *Raphanobrassica* a hybrid between *Brassica oleracea* (2n = 18) and *Raphanus sativus* (2n = 20), has 2n = 38 chromosomes and was artificially synthesized by Karpechenko (1927, 1928). *Galeopsis tetrahit* (2n = 32) is a hybrid between *Galeopsis speciosa* (2n = 16) and *G. pubescens* (2n=16).

If the chromosome numbers found within a group bear no simple numerical relationship to each other, then the series is termed as an aneuploid series or simply aneuploidy. Here two distinct phenomena occur. (1) increase or decrease in the number of chromosomes whereby the same genetic material becomes distributed in a different number leading to a change in basic number: (2) the basic number is unaltered but the genetic material present is changed through the addition or loss of chromosomes.

In nature, a large number of plant groups are known to exhibit aneuploidy. Various species of *Carex* (Cyperaceae) show a wide range of chromosome numbers from n = 6 to 112 with multiples of 5, 6, 7 and 8. The genus *Iseilema* also shows aneuploidy. *Iseilema anthephoroides* is having n = 5, *I. venkateswarlui* n = 4 and *I. lubbardii* n = 3 (Satyavathi and Murthy, 1974). In the genus *Vicia* there are chromosome numbers of 2n = 10, 12, 14, 24 and 28 obviously consisting of aneuploids clustered around the diploid and tetraploid levels, and in *Crepis* 2n = 6, 8, 10, 12, 14, 16, 18, 22, 24, 42, 44, 66, 88 (and some other numbers), with less obvious clustering.

Cytological studies, especially the Karyotype data, have been shown to be useful in taxonomy at all levels up to that of family and even above, but most involve species and groups of species.

Raven and Kyhos (1965) have shown Karyotype affinities for the closely related Magnoliaceae, Himantadraceae and Degeneriaceae and also for Schisandraeae and Illiciaceae while there is marked difference in chromosome numbers of Lauraceae (x = 12) and Hernandiaceae (n = 20, 40). Wiens (1975) has provided cytological evidence, for separation of Viscaceae from Loranthaceae. Various members of Loranthaceae are unified by a single basic chromosome number x = 9 while Viscaceae are characterized by a series of aneuploid numbers ranging from 10 to 14. In most classical systems of classification the genus *Paeonia* is kept in the tribe Paeonieae of the Ranunculaceae. Gregory (1941) noticed that the basic chromosome number in *Paeonia* is five in contrast to seven, eight or nine in the Ranunculaceae, hence advocated its separation into a separate family Paeoniaceae.

The classical example of the use of chromosomal characters in the delimitation of genera can be seen in *Crepis, Cymboseris* and *Youngia* of the Asteraceae (Babcock, 1947). In *Crepis* various species have x = 7, 6, 5, 4 or 3 chromosomes while the basic numbers of other related genera are either x = 8 or x = 9.

Cytological data has been shown to be useful in the taxonomy of *Chlorophytum* (Liliaceae) by Naik (1977). Three species namely *C. bharuchae, C. glaucum* and *C. glaucoides* are morphologically rather difficult to distinguish from one another but *C. bharuchae* has 2n = 16 chromosomes while the other two have 2n = 42. Further, species with 2n = 42 differ, in their karyomorphology.

Chromosome Size and Structure

The most commonly utilized aspect of chromosome structure is the position of centromeres, *i.e.* the arm length ratio of each chromosome in the genome. Chromosomes with centromeres near the middle are known as metacentric, those with centromere near one end are known as acrocentric while the ones with a truly terminal cetromere are known as telocentric.

In many genera and families of flowering plants, conspicuous differences in appearances of Karyotype have been found in species having same chromosome number. In some instances these differences follow definite trends associated with trends of morphological specialization. In the primitive genus *Helloborus* (Ranunculaceae), for example the chromosomes in the Karyotype differ little from each other in size and most of them are V-shaped with median or submedian centromere. Karyotype of this nature are the most common ones. Levizky (1931) considered them as generalized types. Various specialized types have been considered to have been derived from these. Thus, the chromosomes with uniform length may gradually differ in size and shape from V-Shaped (metacentric) to L-shaped (submetacentric) to J-shaped (acrocentric) and ultimately to rod-shaped or I-shaped (telocentric) and become asymmetrical. In the advanced genera *Aconitum* and *Delphinium* (Ranunculaceae), for example, the flowers are zygomorphic with the largest number of J-shaped chromosomes.

Less commonly, the reverse trends of specialization *i.e.,* from telocentric chromosomes to metacentric ones have been demonstrated in certain members of the Commelinaceae (Jones, 1970).

Chromosome structure, together with chromosome size and number have been found extremely useful at all levels of the taxonomic hierarchy (Jackson, 1971). The best known use of the importance of chromosome size is in the Monocotyledonous genera *Yucca, Agave* and relatives. *Yucca* and some others have flowers with a superior ovary, while *Agave* and others have an inferior ovary. For this reason the former were previously placed in Liliaceae and the latter in the Amaryllidaceae. Because of the great overall similarity in these genera they are now usually segregated from their respective families and placed in the Agavaceae. This classification is strongly supported by evidence obtained in the 1930s when it was found that the Agavaceae are characterized by a very distinctive bimodal karyotype consisting of five large chromosomes and 25 small ones (McKelvey and Sax, 1933). This finding, indeed, supported Hutchinson's realignment of these families and his separation of both these genera into a distinct family, Agavaceae.

In the genus *Crepis* the use of chromosome number and morphology has solved many taxonomic problems at the generic, sectional and specific levels and show how many species are evolutionary inter-related (Babcock, 1947).

Chromosome Behaviour

The behaviour of chromosomes during pairing and subsequent separation is some times useful in studying the interrelationships of taxa. Meiosis itself is sometimes provide taxonomic information. For example in Juncaceae and Cyperaceae which have small chromosomes with unlocalized centromeres is considered to be inverted *i.e.,* equatorial division precedes the reductional division instead of vice versa. The phenomenon is not always correlated with unlocalized centomeres and it therefore provide evidence for the close relationship of the above two families.

Many peculiarities of meiotic behavior owe their origin to existence of heterozygosity, so that one is observing pairing of unlike genomes. Some species are known to be consistently heterozygous for certain translocations, which give rise to multivalent formation at meiosis. In the genus *Oenothera* all the species are diploid with 2n = 14. Many of these exhibit normal meiosis, but in the subgenus *Oenothera* the species are heterozygous for translocations involving varying numbers of chromosomes, so that univalents of varying sizes arise at meiosis (Cleland, 1972). In the *O.biennis* group, for example, *O. biennis* itself exhibits a ring of six and ring of eight chromosomes, *O. erythrosephala* exhibits a ring of 12 chromosomes and a bivalent and *O. stringosa* exhibits a single ring of 14 chromosomes.

Banding Patterns

In recent years our ability to distinguish morphologically between chromosomes, atleast in favourable material has been greatly enhanced by new staining techiques using Giesma and fluorochrome dyes which stain chromosomes in a consistent banding pattern instead of with a uniform intensity, as is the case with the usual basic fuchsin (Fuelgen reagent). Schweizer and Ehrendorfer (1976) and Ehrendorfer *et al.* (1977) have applied this technique in the systematics of the genus *Anacyclus* and have found good correspondence with the results obtained by morphological and phytochemical approaches. Their data also correspond well with phylogenetic conclusions derived from other fields of study and suggest a more primitive position for perennial taxa of *Anacyclus* than for the annual ones. *A. officinarum* was shown to be probably a hybrid between *A. pyrethrum* and another annual species *A. radiatus.*

Chemotaxonomy

The subject of chemotaxonomy is concerned with the application of chemical characters to problems of classification and phylogeny. This rapidly expanding discipline of plant taxonomy has been variously called as chemotaxonomy, chemosystematics, biochemical systematics or phytochemistry. Chemotaxonomy has made rapid progress in the last 40 years because of new instruments and newer techniques. Chromatography and elctrophoresis have made the analysis of plants much quicker and simpler. Botanists of late have come to the conclusion that evidence for discussing the relationships and phylogeny must be taken from as many sources as possible. As more comprehensive survey of phytochemistry has become available, taxonomists have shown a growing interest in the application of chemical characters to taxonomic problems.

The earliest attempts to correlate chemistry with the phylogenetic level of development was that of Abbott (1886). According to her the Saponin containing plants occupied the middle level of Hackel's scheme of plant evolution. Greshof (1909) suggested that chemical characters should be included in natural classification. He was also of the opinion that every description of a new genus or species should include a short chemical description of that taxon.

The first successful attempt to combine chemical and morphological evidence in the study of a single genus was the work of Baker and Smith (1920) on the essential oils of *Eucalyptus.* They suggested that the level of relationship should be reflected in chemical similarities (primitive plant chemically primitive). According to the morphological and chemical data collected on 176 *Eucalyptus* species, they divided the genus into three groups differing in both morphological structure and chemical constituents. They concluded that primitive species are those which have feather-veined leaves and high Pinene (I) content in their essential oils (terpenes) while more advanced types have intermediate venation and contain Pinene and Cineole (II). According to their system, the most advanced taxa show butterfly wing venation and contain oils with Phellandrene (III), Piperitone (IV), and Geranyl acetate (V). The work of Baker and Smith was the first comprehensive chemotaxonomic-morphological study of a complex genus, and in their conclusions have been a significant contribution to the taxonomy of *Eucalyptus*.

Considering that the modern development of the topic can be dated quite precisely from the year 1963, the progress since that time has been astonishingly rapid. 1963 saw the publication of the first textbook on the subject by Alston and Turner (1963), the publication of the Proceedings of the First International Meeting on the topic (Swain, 1963) and the launching of Hegnauer's seven volume encyclopaedic series on Plant Chemotaxonomy (Hegnauer, 1962-1974). Later a four volume Guide to Flower Plant Chemotaxonomy by Gibbs (1974) has been published.

Compounds Useful in Plant Taxonomy

Although in theory all the chemical constituents of a plant are potentially valuable to a taxonomist, in practice some sorts of molecules are far more valuable than others. Apart from inorganic compounds, which are of relatively little use, three broad categories of compounds can be recognized: Primary metabolites, secondary metabolites and semantides.

Primary metabolites are parts of vital metabolic pathways and most of them are of universal occurrence, or atleast occur in a very wide range of plants. Aconitic acid (first isolated from *Aconitum*) or citric acid (from *Citrus*), for example, participate in the Krebs (tricarboxylic acid) cycle and are present in all aerobic organisms; the presence or absence of such compounds is therefore not of much systematic value. The same is true of the 22 or so amino acids which are known to be constituents of plant proteins or any of the sugars which figure in the photosynthetic carbon cycle and so on.

In some cases, however, the quantities of such metabolites vary considerably between taxa and this in itself can be taxonomically useful. For example, taxa in which universally occurring substances were first detected (such as the two above) often possess particularly large quantities of the molecules concerned well above the amounts which participate in the essential metabolic pathways often as food-storage materials. Some times such compounds are stored in a different form from that in which they are metabolized, for example sedoheptulose, a sugar constituting the carbohydrate food reserve of the genus *Sedum*, which as sedoheptulose diphosphate is part of the photosynthetic carbon cycle.

Secondary metabolites (or secondary plant products) perform non-vital (or at least non-universally vital) functions and are therefore less widespread in plants. It is of course this restricted occurrence among plants, which renders them valuable as taxonomic information. The most well known groups of compound which have been utilized in this way include alkaloids, phenolics, glucosinolates, amino acids, terpenoids, oils and waxes and carbohydrates.

Secondary metabolites (or secondary plant product) perform non-vital (or at least non-universally vital) functions and are therefore less widespread in plants, It is of course this restricted occurrence among plants which renders them valuable as taxonomic information. The most well-known groups of compounds which have been utilized in this way include alkaloids, phenolics, glucosinolates, amino acids, terpenoids, oils and waxes and carbohydrates.

Secondary plant products are largely waste substances, food stores, pigments, poisons, scents, structural units or water repellents etc. In many cases they obviously do have an essential function but of a general nature so that the precise molecular configuration of the compound is not vital. Thus a yellow pigment with an absorption maximum at 477 nm (which presumably defines its function) might be a betalin or anthocyanin and a poison might be an alkaloid or a glycoside.

Semantides are the information-carrying molecules. DNA is primary semantide, RNA a secondary semantide and proteins are tertiary semantides, following from the sequential transfer of the genetic code from the primary genetic information (DNA). In theory, the sequences nucleotides and amino-acids in these substances should provide all the taxonomic information necessary for classification and offer an alternative to the study of secondary metabolites, cytology, morphology and anatomy etc., for the latter are nearly manifestations of the former. However, in practice there are great difficulties in the gathering of the sequence data and the results frequently present puzzling anomalies. Sometimes the semantides, together with the larger polysaccharides are known as macro-molecules and the primary and secondary metobolites as micromolecules.

Use of Chemical Characters in Plant Taxonomy

In Table 9.5 some major grouping of plant natural products are listed along with some comment on their botanical distribution.

Table 9.5: A Summary of Some Groupings of Natural Products, which have been Used in Chemotaxonomic Correlations including some Comments on the Botanical Distribution of these Constituents (Radford *et al.*, 1976)

Compound Grouping	*Comments in Botanical Distribution*	*Citation*
Terpenoids	Widely distributed	Weissmann (1966, see: Swain)
Monoterpenes	Widely distributed, of Taxonomic value primarily below the generic level	Numerous reports, see Mirov (1967) or others
Sesquiterpenoids	Rather wide distribution, but particularly important and useful in the taxonomy of the Compositae	Herz (1968, see: Mabry *et al.*) Herout and Sorn (1969, see Harborne and Swain)
Asperulosides and Aucubins (Iridoid glycosides)	Rubiaceae, Scrophulariaceae, and related families, Plantaginaceae, Cornaceae	Bate-Smith and Swain (1966, see: Swain) Kooiman (1969, 1970) and others
Ranunculins	Found only in the Ranunculaceae	Bate-Smith and Swain (1966, see: Swain) Kooiman (1969, 1970)
Cyanogenic compounds	Ranunculaceae and other families	Ruijgrok (1966, see: Swain)

Table 9.5–*Contd...*

Compound Grouping	*Comments in Botanical Distribution*	*Citation*
Polyalcohols	Qidely distributed but have taxonomic potential	Plouvier (1963, see: Swain)
Sulfur compounds	A chemically diverse grouping of compounds which in their various forms are widely distributed, the iso-thiocyanate producing glycosides are characteristic of families in the Rhoedales	Kjaer (1966, see: Swain)
Amino acids	Liliaceae and related (non-protein) Families, Leguminosae (particularly Papilionatae, and other groups)	Bell (1966, see: Swain)
Alkaloids	A chemically and biosynthetically diverse group of compounds rarely found in lower vascular plants and irregularly distributed among the angiosperms; highly useful in the taxonomy of some groups	Hegnauer (1966, see: Swain) Schafer (1964) Price (1963, see: Swain)
Betacyanins and betaxanthins	Centrospermae ("Betanales")	Mabry (1966, see: Swain)
Alkanes (fatty acids and waxes)	Widely distributed, possibly of use in classification below the genus level, and of use in organic geochemistry	Shorland (1963, see: Swain)
Fatty acid epoxides	found in seven families	Bu'lock (1966, see: Swain)
Acetylenes	Distributed in the basidiomycetes and atleast 13 angiospermous families, of particular use in the link between the Compositae and the Umbelliflorae	Bu'lock (1966, see: Swain)
Assorted compounds	Ferns	Berti and Bottari (1968, see: Reinhold and Liwschitz)
Diterpenes	Rather wide distribution in the seed plants, taxonomically useful in specific groups such as the conifers	Erdtman (1968, see Mabry *et al.*), Ponsinet *et al.*, 1968, see: Mabry *et al.*)
Triterpenes	Wide distribution in living organisms, taxonomically useful in several Angiosperm families such as the Cucurbitaceae, and the fungi	Ponsinet *et al.* (1968, see Mabry *et al.*)
Carotenoids	The universally distributed photosynthetic carotenoids useful in algal classification; the non-photosynthetic carotenoids in fruits of possible limited value as taxonomic markers in the angiosperms	Goodwin (1966, see: Swain)

Contd...

Table 9.5–*Contd...*

Compound Grouping	*Comments in Botanical Distribution*	*Citation*
Flavonoids	Very large number of diverse compounds found throughout the vascular plants and in nearly, if not all, angiosperms, of great level and of possible use in the classification of higher categories	Bate-Smith (1963, see: Swain); Harborne (1966, see: Swain); Wagner (1966, see Swain); Williams (1966, see Swain); Alston (1968, see Mabry *et al.*) (extensive literature)
Lignins and lignans	Useful in the classification of higher categories	Erdtman (1968, see: Mabry *et al.*)
Quinones	Widely distributed among living organisms, but compounds of this type with limited distribution have potential value in the classification of a number of angiosperms families	Mathis (1966, see: Swain)
Polysaccharides	Universally distributed, probably useful in the classification of higher categories - particularly among the algae, comparative data mostly lacking	Percival (1966, see: Swain)
Plant glycosides	Great chemical variation in the non-sugar portion of the molecule, varyingly of great usefulness in taxonomy	Paris (1963, see Swain)
Assorted compounds	Lichens	Huneck (1968, see Reinhold and Liwschitz)
Wide range of chemical approaches	Umbelliferae	Heywood (1971)
Wide range of chemical approaches	Leguminosae	Harborne *et al.* (1971)

The special value of the chemotaxonomic approach can be seen when chemical characters correlate well with data obtainable from other sources. To take just one example, the fact that there is chemical discontinuity among the families of the order Rhoedales as defined by Wettstein (1935) (Table 9.6) has been used by modern taxonomists as a reason for dividing this assemblage of six or seven families into two orders, the Capparales and the Papaverales (see Cronquist, 1968, 1981). Other reasons for making this split lie in the differences in anatomy (presence or absence of latex system) and in morphology (position of stamens). This new division is also suggested by recent serological and palynological studies of the relevant families.

It should be pointed out in this case that there are many morphological similarities between the separated groups; for example common floral features (*e.g.*, perfect hypogynous flowers, compound ovaries with parietal placentation). Thus, the additional chemical evidence may have been crucially decisive in the acceptance of this newer grouping. In this instance, the chemical data are unusually good in the sense that relatively wide chemical surveys have been conducted in the relevant families. Not only have over 300 crucifer species have been examined for glucosinolates with positive results (Kjaer, 1976) but also all Papaveraceae studied have been shown to have complex mixture of alkaloids in their tissues (Manske, 1966).

Table 9.6: Revised Classification of Families of the Order Rhoedales (Sensu Wettstein)

Families	*Chemical characters*	*Biological characters*
CAPPARALES		
Brassicaceae Capparaceae Resedaceae Moringaceae	Glucosinolates (mustard oil glycosides) universal; isoquinoline alkaloids absent	Centrifugal stamens; no latex system
PAPAVERALES		
Papaveraceae Famariaceae	Glucosinolates absent; isoquinoline universal	Centripetal stamens; well developed latex system

Examples from Secondary Metabolites

Of all the groups of secondary metabolites used by chemotaxonomy, probably none has provided more taxonomic data than phenolic compounds. Plant phenolics are so numerous that detailed reviews on these compounds have been written by Bate-smith (1962), Harborne (1964, 1967) and Ribereau Gayon (1972).

The phenolic molecules most often studied in chemotaxonomy fall into a general class known as flavonoids. They are the largest group of naturally occurring phenols. These are further classified into flavones, flavanones, isoflavonoids, flavanols, anthocyanidins etc.

One of the best examples of the taxonomic value of secondary metabolites concerns flower pigments. The flower pigments, which are usually anthocyanins and anthoxanthins, vary greatly and have been shown to be under genetic control. But at the same time all variation in pigment production is not genetic in origin.

Bate-Smith (1948, 1958, 1962) has reviewed the early application of evidence from phenolics to taxonomic problems. According to him, leuco-anthocyanins are generally present in woody plants. The herbaceous Primulaceae, however, are exceptionally rich in leucoanthocyanins, and some of the woody members of Oleaceae and Scrophulariaceae lack them completely. Phenolic characters have been used by Bate-Smith (1958) in the rearrangement of the species of *Iris*. Williams (1982) pointed out chemical evidence from the flavonoids in the classification of *Malus* species. On the basis of flavonoid patterns. McClure and Alston (1966) identified highly problematic taxa of Lemnaceae a family in which morphological and cytological characters were observed to be inadequate in distinguishing different taxa of *Spirodela, Lemna, Wolfiella* and *Wolffia*. Williams (1976) investigated leaf flavonoids of Liliaceae, Juncaceae, Cyperaceae and Poaceae and supported the view that all these four families have arisen from Liliaceous ancestors. Williams *et al.* (1981) observed Araceae to be unusual amongst Monocots because of their simple and relatively uniform flavonoid profile. No one subfamily was clearly distinguishable and it was only at the tribal level where some significant taxonomic patterns were observed. Their observations indicated that this family has not been derived from Liliaceous stock since flavone C-glycosides, the characteristic constituents of the Aroids, were rarely encountered in the present day Liliaceous taxa. More over, they were observed to be totally absent from the tribe Aspidistreae from which the origin of this family has been suggested by Hutchinson (1959).

Chemistry is useful to taxonomy not only at generic level but also in family assignments. There are still a considerable number of usually small angiosperm families whose affinities are not clear and which are difficult to place with certainly into an ordered system of classification. One such family is the Julianaceae the taxonomic relationships of which have been the subject of controversy for atleast a century. Based on wood anatomy and morphological features associated with wind pollination, this family has variously been placed in three different associations, near either the Anacardiaceae, or the Burseraceae or the Juglandaceae.

Chemical analysis of the heartwood flavanoids by Young (1976) provides clear-cut data in support of the first of these associations. Thus the heartwood of *Amphipterygium adstrigens* (Julianaceae) yielded sixteen flavonoids, seven of which were 5-deoxy-flavonoids. These rare substances occur quite characteristically in the Anacardiaceae but are not known in either the Burseraceae or the Juglandaceae. Indeed, all seven of these 5-deoxyflavonoids have been recorded widely in the tribe Rhoeae of the Anacardiaceae and one of them, the aurone rengasin, is only known otherwise from this family. In contrast, no evidence could be found in *Amphipterygium* of Juglone, which is a characteristic quinone of Juglandaceae.

By using the new flavonoid data together with a reappraisal of the anatomical characters, Young (1976) decided to reduce the Julianaceae to subtribal rank and include it in the Anacardiaceae as the subtribe Julianiinae of the tribe Rhoeae. This decision has been accepted by Thorne (1981) in his latest revision of Angiosperm classification and is also in line with recent serological investigations of these and related families (Petersen and Fairbrothers, 1979).

Analysis of flavonoid, carotenoid and quinonoid pigments in plants of the Gesneriaceae have shown significant correlations with the sub-family division of Burtt, based on geography and anisocotyly, while not fitting in with the earlier division on von Fritsch, based on position of ovary (Harborne, 1967). Similarly phytochemical surveys in the tribe Anthemideae (Asteraceae) for both flavonoids and polyacetylenes have indicated that the morphological splitting of the *Chrysanthemum* complex *sensu lato* into smaller generic groups is justified on chemical grounds (Bohlman *et al.,* 1964; Harborne *et al.,* 1970). In the tribe Genisteae of the Fabaceae a survey of the isoflavones, proanthocyanidins and flavones in the leaf provided results which were closely in accord with Rothmaler's treatment of the genera but were at variance with Hutchinson's groupings of the same taxa (Harborne, 1969).

Considering the taxonomic importance of instances of compounds with unique distribution in plants, the study of the "nitrogenous anthocyanins" is perhaps the best known example. The Betalins, as they are now known, include the red to violet betacyanins and the yellow betaxanthins.

Betalins have been found in nine families of Angiosperms, all members of the order Centrospermae. These are: Aizoaceae, Amaranthaceae, Basellaceae, Cactaceae, Chenopodiaceae, Didieraceae, Nyctaginaceae, Phytolaccaceae and Portulacaceae. Engler and Prantl (1887-1915) kept the Cactaceae in a separate order. But is has now been found that the Cactaceae also shows Betalins and hence they should be included in the Centrospermae. Two families – Molluginaceae and Caryophyllaceae – which are often included in the Centrospermae do not possess Betalins and instead show Anthocyanins.

According to Mabry *et al.* (1976) these two families should be kept in a separate suborder Caryophyllinae in the order Caryophyllales. These should be included in Centrospermae on ultrastructural evidence as they also show P III subtype sieve element plastids (table 9.7).

The genus *Gisekia,* traditionally assigned to the Molluginaceae because of morphological similarities, is anamolous in that family because it has betalins instead of anthocyanins. The genus has recently been transferred to the Phytolaccaceae by Takhtajan (1980), where it seems to belong more naturally than in any other family of Centrospermae.

Table 9.7: Classification of the Order Centrospermae (Caryophyllales) Varying According to the Use of Evidence from Pigments

Structural Classification Engler and Prantl (1897-1915)	*Chemical Classification*	*Compromise Classification*
Centrospermae	*Chenopodiales*	*Order : Caryophyllales*
		Suborder : Chenopodinae
Aizoaceae	Aizoaceae	Aizoaceae
Amaranthaceae	Amaranthaceae	Amaranthaceae
Basellaceae	Basellaceae	Basellaceae
Caryophyllaceae	Cactaceae	Cactaceae
Chenopodiaceae	Chenopodiaceae	Chenopodiaceae
Didieraceae	Didieraceae	Didieraceae
Molluginaceae	Nyctaginaceae	Nyctaginaceae
Nyctaginaceae	Phytolaccaceae	Phytolaccaceae
Phytolaccaceae	Portulacaceae	Portulacaceae
Portulacaceae		
Cactales	*Caryophyllales*	*Suborder : Caryophyllinae*
Cactaceae	Caryophyllaceae	Caryophyllaceae
	Molluginaceae	Molluginaceae

Cyanogenic Compounds

Hegnauer (1977) defined the term cyanogenesis as the ability of certain plants to release hydrocyanic acid after injury of cells. Usually cyanophoric plants contain one or several cyanogenic glycosides. Since 1801, when Bohm first detected HCN and amygdalin (the first plant glycoside) in seeds of *Prunus amygdalus* (Rosaceae) many Rosaceous and other plants have been shown to be cyanophoric. Till date about 2,056 species of vascular plants are known to be cyanophoric. In Angiosperms they occur erratically. Cyanogenic taxa are relatively frequent in Araceae, Juncaceae, Juncaginaceae, Gramineae and Scheuchzeriaceae among Monocotyledons and several Dicotyledonous families.

Cyanogenesis, if adequately and carefully used as a character in plant systematics, may prove to be of considerable value in the classification of plants even at the higher systematic levels (Hegnauer, 1977). The distribution of individual cyanogenic compounds in the tracheophytes supports the theory that the Liliopsida (Monocotyledons) evolved from ancestors resembling present day Magnoliidae because cyanogenesis proceeds in exactly the same way in both these groups, with only tyrosine as the precursor (Hegnauer, 1973, van Valen, 1978a).

The subfamilies of Poaceae are characterized by three different tyrosine derived cyanogenic compounds, namely Dhurrin (Andropogonoideae), Triglochinin (Festucoideae and Eragrostioideae) and Taxiphyllin (Bambusoideae) (van Valen, 1978b).

The facts known today suggest that more than one biosynthetic group of cyanophoric compounds occur only in very large genera or families belonging to Dilleniidae, Rosidae and Asteridae. In these taxa, accumulation of Cyanogenic compounds is taxonomically significant at the infra-familiar level. Linamarin for example, is a character of many Loteae, Trifolieae and Phaseoleae in Fabaceae and Calenduleae in Asteraceae.

In India several botanists have contributed to the Chemotaxonomy of plants. Notable among them are S.D. Sabnis from Baroda, L.L. Narayana from Warangal and M. Radhadrishnaiah, from Hyderabad.

Nageswar *et al.* (1984) has studied the distribution of different chemical compounds in *Caesalpinia* and gave a key for identification of the species on chemical grounds.

1. Juglones present .... *C. sappan*
1. Juglones absent:
    2. Aucubin compounds present .... *C. coriaria*
    2. Aucubin compounds absent:
        3. Catechol tannins present:
            4. Methylene dioxy compounds present .... *C. pulcherrima* (red)
            4. Methylene dioxy compounds absent.......... *C. pulcherrima*(Yellow)

3. Catechol tannins absent:
    5. Leucoanthocyanins Present:
        6. Anthraquinones Present .... *C. gledeschiodes*
        6. Anthraquinones absent:
            7. Tannins present .... *C. cacalace*
            7. Tannins absent .... *C. sepiaria*
    5. Leucoanthocyanins absent:
        8. Syringaldehyde doubtful .... *C. bonducella*
        8. Syringaldehyde absent:
        9. Steroids present .... *C. digyna*
        9. Steroids absent:
            10. Tannins present .... *C. ferrae*
            10. Tannins absent .... *C. tortuosa*

Radhakrishnaiah *et al.* (1982) on phytochemical grounds supported the separation of *Trapa* from Onagraceae. The phytochemical studies by Radhakrishnaiah *et al.* (1984) lend support to the close kinship among the genera *Sparganium, Typha* and *Pandanus* and their retention under one Englerian taxon Pandanales and negate their segregation. Anuradha *et al.* (1988) studied the chemotaxonomy of Capparaceae. Their study does not lend support the separation of Cleomaceae from Capparidae. The chemical data given by Rajendra Prasad *et al.* (1986) do not warrant the creation of Mimosaceae as an independent family.

Examples from Semantides

Serology and Taxonomy

Boyden (1964) defined Serology as that portion of Biology which is concerned with the nature and interactions of antigenic material and antibodies. The science dealing with immunochemical reactions between, serum antibodies and antigens (Klaus, 1971) has provided valuable taxonomic information because its techniques help to detect homologous proteins. Boyden (1967) and Fairbrothers (1968, 1977) have discussed the importance of this type of study in plant systematics. A protein extract of plant or animal origin (antigen) is injected into the blood stream of an experimental animal, usually a rabbit, to form antibodies. A specific antibody is produced in response to a specific antigen. The serum (termed the antiserum) is then made to react *in vitro* with the antigenic proteins as well as those of other taxa of which the affinities are in question. The amount of precipitation indicates the degree of protein homology and hence is taken as a phylogenetic marker and taxonomic character. Kowarski (1901), Bertarelli (1902) and Magnus (1908) were the first notable serologists. They compared proteins from various grass and legume species showing similarities and differences.

Serological studies using crude plant protein extracts have been widely used in estimating phylogenetic relationships and elucidating the taxonomy of a wide variety of taxa. Thus a serological comparison of major seed proteins from Angiosperms has indicated a close relationship among the Magnoliidae, Hamamelididae and Corniflorae (sensu Dahlgren, 1980) and has refuted the idea of their independent evolution. It has also confirmed the homogeneity of the iridioid – producing Corniflorae and support the inclusion of the Gentinaceae in it (Jensen and Greven, 1984).

Based on serological studies the genus *Liriodendron* had been found to be quite distinct from other members of the family Magnoliaceae, and the genera *Magnolia* and *Michelia* displayed closest affinity within the family (Johnston and Fair brother, 1965). Similar studies have also been useful in estimating the relationship of the Nymphaeaceae and Nelumbonaceae (Simon, 1971), generic kinship in the Caprifoliaceae (Hill-brand and Fairbrothers,1969, 1970) and in the Rubiaceae and related groups (Lee and Fairbrothers, 1978). In the Umbelliferae the findings support the classification of the family into Hydrocotyloideae, Saniculoideae and Apioideae and suggest the Apioideae is more closely related to Saniculoideae than the Hydrocotyloideae (Pickering and Fairbrothers, 1970). Chrispeels and Gartner (1978) found serological evidence for assigning *Phaseolus aureus* and *P.mungo* to the genus *Vigna*. The classification of the Ranunculaceae at tribal and generic levels indicated by serological relationships strongly supported that based on classical evidence, particularly from the chromosomes (Jensen, 1968, 1974). At lower levels Smith (1972) has studied the relationship of annual brome-grasses (*Bromus*) relating immuno-electrophoretic data to those from morphology, cytology and cytogenetics. He was able to throw new light on species problems, diploid-tetraploid relationships and hybrids, and the serological and cytological distinctness of a previously named variety of *Bromus secalinus* led him to recognize it as a new species *B. pseudosecalinus*.

DNA and RNA Hybridization

It has so far not proved possible to sequence DNA or the larger RNA molecules in a routine way and, for this reason nucleic acid chracteristics are inferred by indirect means such as DNA hybridization. In brief, DNA extracted from one organism is treated to convert it to a single stranded polynucleotide chain, and the amount of reassociation (annealing) with similarly treated DNA from another taxon which occurs on mixing the two is taken as a measure of similarity of the nucleotide sequences. Bendich and Botton (1966) discussed the relationships among the Leguminosae as measured by the DNA-Agar technique. In quantitative species comparisons based on DNA hybridization to and for *Secale, Hordeum* and *Triticum* showed, that, of these Cereals, wheat (75 per cent) is more closely related to rye (100 per cent) than is barley (75 per cent)

RNA has been investigated by the equivalent technique of DNA-RNA hybridization, whereby the amount of association occurring between DNA of on organism and a fraction of the RNA (usually ribosomal RNA) of another is taken as a measure of relationship. Mabry (1976) investigated plants of the Centrospermae (Caryophyllales) by this method, and concluded that Caryophyllaceae (which contain anthocyanins instead of betalins) are quite close to the betalin-containing families, but not as close as the latter are to each other.

Amino Acids Sequencing

Amino acids sequencing is a more modern development, aiming to identify pure proteins down to the atomic level. It is now possible to break off the amino acids from the polypeptide chain one by one, identify each in turn chromatographically, and so build up the complete sequence of amino acids step by step. A first step is to break the total polypeptide chain into smaller peptide fragments, each of which is then sequenced separately. In the past these small peptide units were separated by chromatography and/or electrophoresis (fingerprinting), and such relatively crude results themselves proved of taxonomic value. Amino acids sequencing investigates the variation in the precise sequence of amino acids in a single homologous protein (*i.e.* one presumable of monophyletic origin) throughout a range of organisms. This relies upon the fact that a particular protein does not have a single invariable structure, but a good proportion of it may vary without altering its essential functions. In the case of cytochrome C for example (the molecule most used for sequencing purpose), about 79 out of approximately 113 amino-acids vary from species to species, but alteration of even one of the other 34 destroys the functioning of the molecule. Cytochrome C is an ideal molecule as it is relatively small and stable, is coloured and is ubiquitous is aerobic organisms. So far its sequences in atleast 25 species of vascular plants have been determined besides others in algae, animals, fungi and bacteria.

In general, the number of differences is closely parallel to the distance apart of the organisms in traditional classifications, suggesting that the method is broadly reliable (Boulter, 1974 a, b, Boulter *et al.*, 1972). Certain amino acids have very restricted distribution and this feature has been fruitfully exploited to characterize certain subfamilies and genera. Canavanine has been observed only in the taxa of the subfamily Lotoideae of Fabaceae (Turner and Harborne, 1967). Lathyrine has been found to be restricted only to the genus *Lathyrus* (Bell, 1971).

Electrophoresis

Electrophoresis is one the most extensively used techniques in protein investigations. The underlying principle is quite simple, proteins, when subjected to an electrical field in a solution of suitable pH, will migrate because of the presence of ionisable molecules on their surface. Under identical conditions the rate of migration of each protein is constant and hence can be reliable character for the detection of homologous proteins. Separation of proteins can be done either on paper (paper electrophoresis) or on a gel medium (gel electrophoresis). The importance of electrophoretic evidence in plant systematics has been discussed at length by Gottlieb (1977). This has been employed in estimating the parental affinities of the interspecific hybrid *Galeopsis tetrahit* (Houts and Hillebrand, 1976). Natarella and Sink (1975) studied the peroxidases and proteins of four species of the genus *Petunia* (*viz. P. axillaris. P. inflata, P. violacea* and *P. parodii*) and some 11 cultivars of *P. hybrida* by this method. Electrophoretic patterns revealed that all the species are closely related and that *P. inflata*, as well as *P. axillaris* and *P. violacea* is involved in *P. hybrida*. Crawford (1985) has demonstrated that electrophoretic data can be a useful supplement to other data used to infer modes of speciation among plants.

The most valuable electrophoretic results have perhaps been obtained in the cereals related to wheat, in relation to the genomic constitution and ancestry of the tetraploids and hexaploids. For example, Johnson (1972), Johnson and Hall (1965) working on storage proteins, concluded that the hexaploid bread-wheat (*Triticum aestivum*) did indeed contain a sum of the proteins possessed by the diploid species, which had been postulated on morphological and cytological evidence to be ancestral to it.

Chapter 10
Taximetrics

Taximetrics

Plants are classified based on their characters. Taxonomists classified plants giving particular reliance to one character, for example one cotyledon (Monocotyledons), two cotyledons (Dicotyledons) etc. A classification with emphasis on one character is called as *a priori* reasoning.

Adansonian Principles

Michel Adanson (1763) for the first time pointed out that equal weightage should be given for all the characters while classifying plants. These principles are called Adansonian principles. The Adansonian principles have received great support since 1960s and have developed new methods in taxonomy included under a general term Numerical taxonomy.The principles of modern numerical taxonomy developed by Sneath and Sokal (1973) are based on the modern interpretation of the Adansonian principles and as such are termed neo-Adansonian principles. These principles of numerical taxonomy are enumerated below (Singh, 1999)

1. The greater the content of information in the taxa of a classification and the more characters it is based upon, the better a given classification will be.
2. *A priori,* every character is of equal weight in creating natural taxa.
3. Overall similarity between any two entities is a function of their individual similarities in each of the many characters in which they are being compared.
4. Distinct taxa can be recognized because correlations of characters differ in the groups of organisms under study.

5. Phylogenetic inferences can be made from the taxonomic structures of a group and from character correlations, given certain assumptions about evolutionary pathways and mechanisms.
6. Taxonomy is viewed and practiced as an empirical science.
7. Classifications are based on phenetic similarity.

Phenetics and Phyletics

Numerical Taxonomy may be defined as numerical evaluation of the apparent (phenetic) similarity among the groups of organisms and ordering them into higher ranking taxa (Heywood, 1967). Numerical taxonomy is empirical and aims to develop objective methods. They can be tested through repeated experiments for evaluation of taxonomic relationships and in the erection of taxa in hierarchial system of taxonomy. Besides, numerical methods have provided clues in phylogenetic analysis for evaluation of evolutionary rates more accurately.

The branch of botany concerned with numerical approaches towards taxonomy has been variously termed as Numerical Systematics (W.H.Wagner), Mathematical taxonomy (Jardine and Sibson, 1971), Taxometrics (Mayor, 1966), Taximetrics (Roger, 1963) and Numerical Taxonomy (Sneath and Sokal, 1973).

In Numerical Taxonomy first of all individuals are selected and there after their characters are spotted out. On the basis of character analysis the resemblance amongst the individuals are established. These stages are often worked out with the help of computers. The accuracy in the estimation of resemblance between taxonomic groups depends on the appropriateness in character. Choice of character requires a careful approach while working on these lines. The selection of character may be made from either existing literature or original observations from the specimen itself. So far as the character number is concerned there is no limitation but larger the number better is the approach for generalization of the taxa. It is widely regarded that a number of 80-100 characters is desirable in numerical taxonomic studies.

There are two aspects of Numerical taxonomy - the construction of taxonomic groups and their discrimination.

Construction of Taxonomic Groups

For designing natural system of classification, two major bases have been recommended: firstly phenetic and secondly phylogenetic (cladistic).

Phenetic classification is the similarity (relationship) of individual under consideration based on a set of phenotypic (characters related to the appearances) characters. This phenotypic relationship expressed in terms of similarity does not embody any relationship through ancestry.

Cladistic relationship is expressed in terms of correlation amongst individuals with regard to their evolutionary history. The relationship means common ancestry. The word cladistic means study of evolutionary sequence and pathway followed by individuals and the origin of branches from evolutionary tree. The evolutionary tree with anastomosing branches leading to various taxa are termed as cladograms.

Summarizing the Data and Analysis of Relationship and Distance Among the Taxa

The basic unit in Numerical taxonomy is known as operational taxonomic unit (OTU). It can be an individual, species, genus, family, order, or class. When the OTU is supra-individual (above the level of an individual), there should be adequate representation of various polymorphic forms *i.e.*, when genera are compared they should be represented by different species, when families are compared, they should be represented by different genera and so on. It should always be borne in mind, that comparison of OTUs of equal rank is made in Numerical taxonomy.

Unit Characters

The characters employed in Numerical taxonomy are known as Unit characters. A unit character is that one which cannot be sub-divided logically. There are two types of unit characters.

1. Those unit characters which exist in two states. They are called binary or two state characters. This is the simplest form of coding where characters are divided into + and – or as 1 and 0. The positive characters are recorded as + or 1 and negative characters as – or 0. In case the organ possessing a character is missing in the organism, the character must be scored NC, which means No comparison. The example for two state character is the presence or absence of spines.
2. Those unit characters which exist in more than two states are called multistate characters. The multistate characters may be (i) qualitative such as the colour of the flower, the colour of the flower could be in any number of states – white, red, yellow, violet, blue etc. or (ii) quantitative such as the length of the leaf, it could be in any number of states – 1 cm, 2 cm, 3 cm, 4 cm etc. The multistate characters can be converted into two state characters like flowers white versus coloured, leaf long versus short. It may be noted that character is an attribute that exists in abstract form, while the character state is an expression of the character that exists in concrete form. The phyllotaxy is a character and alternate, opposite or whorled forms are character states. There is another term characteristic associated with the above two terms. When a character-state is exclusive to a particular taxon/taxa, it is called characteristic.

The proper selection of characters is a critical point in the application of numerical taxonomy. Certain characters are clearly disqualified for numerical taxonomy and these are listed by Sokal and Sneath (1963) as inadmissible characters. According to these authors it is undesirable to use.

1. Attributes which are not a reflection of the genotypes of the organisms themselves.
2. Any property which is a logical consequence of another, either partly or wholly; and
3. Characters which do not vary within the entire sample of organisms.

To be qualified as an admissible character:

1. It should be inherent in the organism and should neither be susceptible to environmental changes, nor affected by experimental uncertainties,
2. It must be of some diagnostic value,
3. It cannot be sub-divided,
4. The character that is selected must show variation among the taxa under comparison.

Measurement of Resemblances

There are three methods of estimating phenetic resemblance between the taxonomic groups, namely

1. Coefficients of association,
2. Coefficients of correlation and
3. Measurement of taxonomic distance.

Cluster Analysis

Different Operational Taxonomic Units (OTUs) are grouped together on the basis of degree of similarity. These groups of OTUs are termed clusters.

There are several techniques to describe structure in matrices of similarity coefficients. One of the common techniques is the differential shading of the similarity matrix. In this method, similarity coefficient are grouped into five or ten evenly spaced classes. Each of these classes are represented by different degrees of shading in the squares of half matrix. The highest value is generally shown darkest and the lowest value the lightest as in Figure 10.1. Then the half matrix can be seen as a pattern to different shades, limited by diagonal of squares with the darkest shade (Figure 10.2). On rearrangement of the sequence of OTUs, clusters can be more sharply defined as in Figure 10.2C.

The groups of similar organisms organized in this manner are termed phenons. The clusters of phenons are then rearranged in a dendrogram, which summarizes the main features of cluster analysis.

The second way of clustering is by dendrograms. In this case the mutually most similar taxa are paired. The pairs are successively joined by the average similarity. The process is continued till all units have been joined together. It results in a tree like dendrogram with taxa at the tips of branches. The horizontal (phenon line) intersecting the vertical lines of the dendrogram gives number of clusters (Figure 10.1).

The delimitation of phenons is done by drawing a horizontal line across the dendrogram (Figure 10.1) at a similarly value. A line at 75 per cent for example creates five 75-phenons 1; 7; 3, 5, 6; 4, 9, 10, and 2, 8; while that at 80 per cent creates six 80-phenons. Such a dendrogram will have a reference to a given taxon and cannot be transferred to any other study. In the above dendrogram, if the OTUs 1 to 10 had been species, an 80-phenon line could indicate 6 subgenera and a 65-phenon line

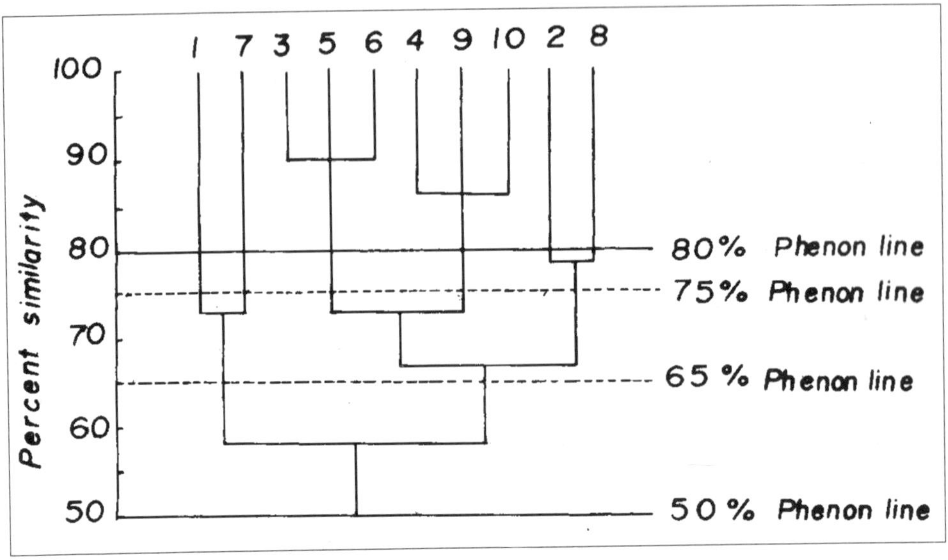

Figure 10.1: Dendrogram ot Show Formation of Phenons

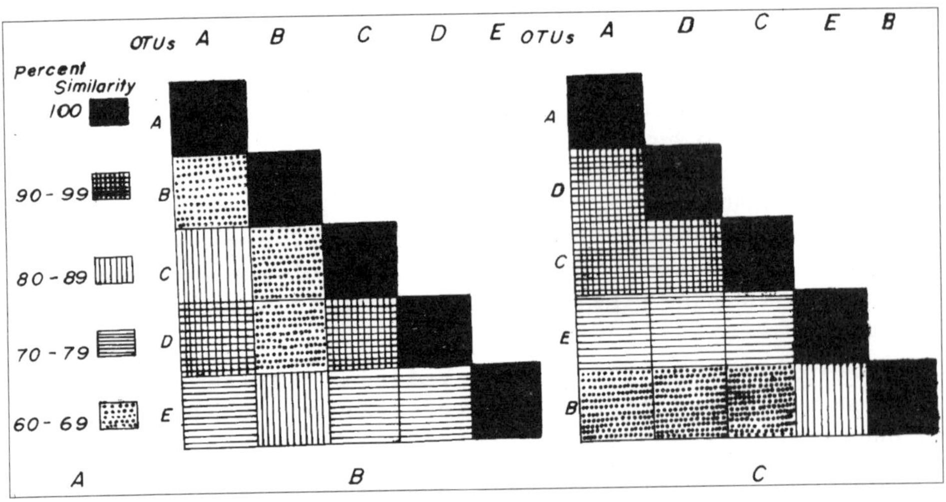

Figure 10.2A-C: Shaded Similarity Matrices

A: Percentage Similarity; B: OTUs Arranged; C: After Rearrangement of OTUs

two genera. It should however be remembered that phenons are arbitrary and relative groups.

Discrimination

If taxonomic groups chosen for the study show overlapping of characters, discrimination should be used to select them. Various techniques, such as discriminant analysis, have been devised for such purposes. The best methods for delimiting taxa are based on the utilization of maximum number of characters with similar weightage given to them.

Nomenclature and Numerical Taxonomy

Modern nomenclature does not concern itself with the problems of delimitation of taxa. It serves only as a reference point to the taxonomic names. The limits are debatable subjective and forever changeable. Numerical taxonomy, on the other hand, is very useful in delimitation of taxa by exact estimation of affinities (although phenetic). Thus, there is no scope for personal opinion or decision of taxonomists. The limits may be objective, utilitarian, permanent and fixed by common consent.

Merits and Demerits of Numerical Taxonomy

According to Davis and Heywood (1963), it would be better to welcome these procedures with caution, since these methods are only an extension of the orthodox procedures. They have raised some doubts. First, the methods will clearly be useful in phenetic classification, not phylogenetic. Similarly, the proponents of "biological" species concept, may not accept the specific limits bound by these methods. Even the practicing taxonomist might use his brain more efficiently than the machine which is fed with non-relevant selection of characters. Character selection is the weak link in this approach. The statistical methods are likely to give less satisfactory solution if characters chosen for comparison are inadequate.

Stearn (1968) indicated that different taxonometric procedures may yield different results. A major difficulty for the beginner is to choose a procedure for his purpose. Another difficulty concerns the number of characters (from 40-100) needed in order to obtain satisfactory results by these mechanical aids. Taxonomists usually manage with far less characters. Further, so far it has not been seen that the results achieved by mechanical means are in any way more acceptable than those visualized by practicing taxonimists. It is desirable to ascertain whether a large number of characters would really give satisfactory results than those using smaller number. Stearn (1964), after applying taxonometric procedures to the Jamaican species of *Columnea* and *Alloplectus* (Gesneriaceae), came to realize that it seemed a pity not to make further use of these. This survey, as Stearn (1968) has concluded, demonstrated the capacity of computer-aided taxonometric methods to build from an assemblage of characters a grouping of species comparable in validity to one made by conscious taxonomic effort. It also indicated that the number of characters used is less important than their range.

Johnson and Holm (1968) after analyzing their data on the genus *Sarcostemma* (Asclepiadaceae) by various taxonometric methods, have concluded that the numerical classification based on correlation coefficients bears closer resemblance to the classical taxonomic classification. However, they expressed their view that thorough analysis of character sets will lead to a better understanding of the process of evolution and the role of environment in determining patterns of variation.

Dale (1968) while presenting the basic procedures which underlie numerical taxonomic methods concluded that any taxonomist proposing to use such methods must be careful in his choice and be wary of what may seem to be unimportant details.

Cullen (1968), while reviewing the botanical problems of numerical taxonomy, extended his welcome to the advance of numerical techniques which according to him, may well provide means of checking and improving classifications by orthodox taxonomists. He has also realised that the numerical classifications are not likely to supplant orthodox ones – they may either confirm them or, if very differtent, exist side by side with them.

Clifford (1970) seems to have used numerical methods for a better classification of the grasses and concluded that there is a greater probability of their evolution from the palms. This conclusion was also drawn by Meeuse (1966) on the basis of the ovarian structure and origin. In both palms and grasses, the ovary has an ecarpellate origin, with greater reduction in the grasses than the palms. Corner (1966) also has indicated that the embryos of *Archantophoenix* possesses a coleorhiza as in grasses, a feature otherwise unknown in palms.

Cladistics

Cladistics (Greek: *Klados* = branch) is a form of biological systematics which classifies living organisms on the basis of shared ancestry. It can be distinguished from other taxonomic systems, such as phenetics, by its focus on evolutionary relationships; while other systems usually use morphological similarities to group similar species into genera, families and other higher level classification, cladistics tries to construct a tree representing the ancestry of organisms and species. Cladistics is also distinguished by its emphasis on objective, quantitative analysis, rather than subjective decisions that some other taxonomic systems rely upon.

Cladistics originated in the work of the German entomologist Willi Hennig, who himself referred to it as phylogenetic systematics; the use of the terms "cladististcs" and "clade" was popularized by other researchers (Hennig, 1966).

Cladistics use cladograms, diagrams which show ancestral relations between organisms, to represent the evolutionary tree of life. Although traditionally such cladograms were generated largely on the basis of morphological characters, DNA and RNA sequencing data and computational phylogenetics are now very commonly used in the generation of cladograms.

Clades

A *clade* is a group consisting of a species (extinct or extant) and all its descendants. In the terms of biological systematics, a clade is a single "branch" on the "tree of life". The idea that such a "natural group" of organisms should be grouped together and given a taxonomic name is central to biological classification. In cladistics (which takes its name from the term), clades are the only acceptable units. The term was coined in 1958 by English biologist Julian Huxley.

A *clade* is termed monophyletic, meaning it contains one ancestor (which can be an organism, a population, or a species) and all its descendants. The term clade refers to the grouping of the ancestor and its living and/or deceased descendants together. The ancestor can be a theoretical or actual species.

Clade Definition

There are three major ways to define a clade for use in a cladistic taxonomy.

- ☆ *Node based*: the last common ancestor of A and B, and all descendants of that ancestor. Crown groups are a type of node-based *clade.*
- ☆ *Branch-based*: the first ancestor of A which is not also an ancestor. (This type of definition was originally called "stem-based", but this was changed to avoid confusion with the term "stem group"). Total groups are a type of branch-based *clade.*
- ☆ *Apomorphy-based*: the first ancestor of A to possess derived trait M homologously (that is, synapomorphically) with that trait in A, and all descendents of that ancestor. The process of identifying and naming groups based on apomorphies is the method that most resemble classical systematics, with the exception that cladistic taxa always denote a *clade.*

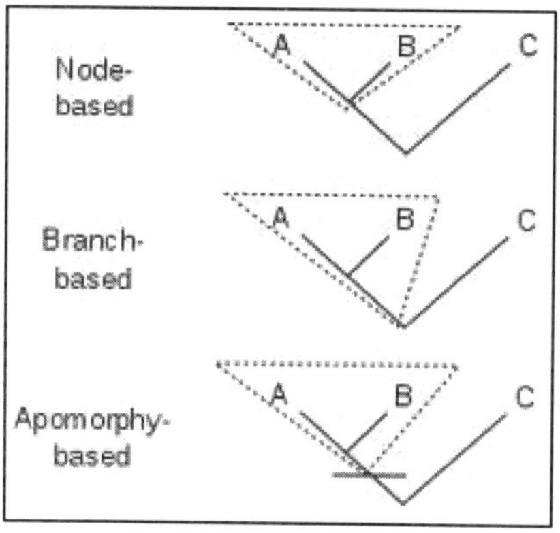

Figure 10.3: Cladograms

In Linnaean taxonomy, *clades* are defined by a set of traits (apomorphies) unique to the group. This system is basically similar to the apomorphy-based clades of phylogenetic nomenclature. The difference is one of weight: While phylogenetic nomenclature bases the group on an ancestor with a certain trait, Linnaean taxonomy uses the traits themselves to define the group.

Clades as Constructs

In cladistics, the clade is a hypothetical construct based on experimental data. Clades are found using multiple (sometimes hundreds) of traits from a number of species (or specimens) and analysing them statistically to find the most likely phylogenetic tree for the group. Although similar in some ways to a biological classification of species, the method is statistical and thus directly open to scrutiny and reinterpretation. With changing phylogenetic analysis, the actual content of any defined clade involved may change as well. Although taxonomists use clades as a tool in classification where feasible, the taxonomic "tree of life" is not the same as the cladistic. The traditional genus, family, etc. names are not necessarily *clades*; though they will often be.

Clade Names

In Linnaean systematics, the various groups are ordered into a series of taxonomic ranks (the familiar order, family etc.). These ranks will by convention dictate the ending to names for some groups. *Clades* do not by their nature fit this scheme, and no such restriction exists as to their names in cladistics. There is however a convention for naming more or less inclusive groups, which are given prefixes like *crown-* or *pan.*

The idea of a "*clade*" did not exist in pre-Darwinian Linnaean taxonomy, which was based by necessity only on internal or external morphological similarities between organisms – although as it happens, many of the better known animal groups in Linnaeus' original Systema Naturae (notably among the vertebrate groups) do represent clades. The phenomenon of convergent evolution is however responsible for many cases where there are misleading similarities in the morphology of groups that evolved from different lineages.

With the publication of Darwin's theory of evolution in 1859, taxonomy gained a theoretical basis, and the idea was born that groups used in a system of classification should represent branches on the evolutionary tree of life. In the century and a half since then, taxonomists have worked to make the taxonomic system reflect evolution. However, partly because the Tree of Life branches rather unevenly, the hierarchy of the Linnaean system does not always lend itself well to representing clades. The result is that when it comes to naming, cladistics and Linnaean taxonomy are not always compatible. In particular, higher level taxa in Linnaean taxonomy often represent evolutionary grades rather than *clades*, resulting in groups made up of clades where one or two sub-branches have been excluded.

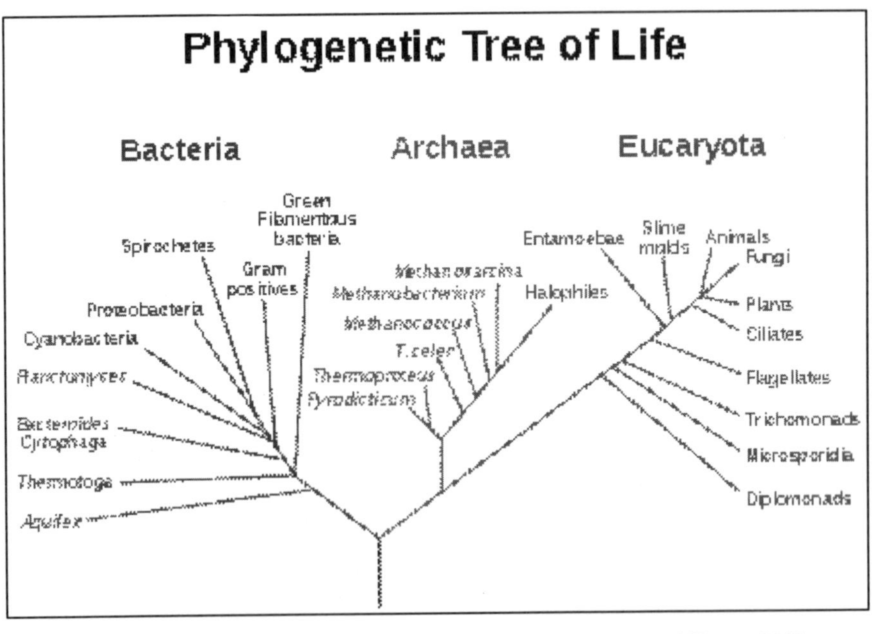

Figure 10.4: A Highly Resolved, Automatically Generated Tree of Life, Based on Completely Sequenced Genomes

In phylogenetic nomenclature, clades can be nested at any level, and do not have to be slotted into a small number of ranks in an overall hierarchy. In contrast, the Linnaean units of "order", "class" etc. must be used when naming a new taxon. As there are only seven formal levels to the Linnaean system (species being the lowest), only a finite number of sub- and super-units can be created. In order to be able to use the full complexity of taxonomic trees (cladograms) in an area with which they are very familiar, some researchers have opted to dispense with ranks all together, instead using *clade* names without Linnaean ranks. The reason for preferring one system over the other is partly one of application: cladistic trees give details, suitable for specialists; the Linnaean system gives a well ordered overview, at the expense of details of the phylogenetic tree.

A *dendrogram* is a broad term for the diagrammatic representation of a phylogenetic tree.

A *cladogram* is a phylogenetic tree formed using cladistic methods. This type of tree only represents a branching pattern, *i.e.*, its branch lengths do not represent time or relative amount of character change.

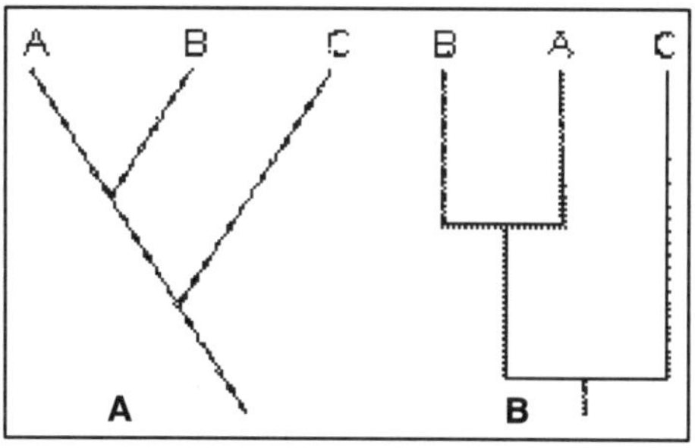

Figure 10.5: A. Cladogram, B. Dendrogram

A *phylogram* is a phylogenetic tree that has branch lengths proportional to the amount of character change.

A *chronogram* is a phylogenetic tree that explicitly represents evolutionary time through its branch lengths

An important idea here is that this cladogram is an evolutionary hypothesis. It is falsifiable.

Used Terms

Three main types of groups can be identified in *cladograms*:

Monophyletic groups are groups consisting only organisms descended from a particular organism.

A paraphyletic group is a subgroup of a monophyletic group, which consists of some but not all the descendents of a single ancestor.

A polyphyletic group consists of organisms from two different monophyletic groups.

The following terms are used to identify shared or distinct characters amongst groups:

- ✰ An ancestral state or plesiomorphy (meaning "close form") is a characteristic that is present at the base of the tree.
- ✰ A derived state or apomorphy (meaning "separate form") is a characteristic which is believed to have evolved within the tree.

Apomorphy and Plesiomorphy

Relationships between taxa are defined in terms of common ancestral species. Two species are more closely related to one another than to a third species if they share a more recent common ancestor that either does with the third. These two taxa are considered as sister groups.

Apomorphy

A derived or specialized character. A trait which characterizes an ancestral species and its descendents is called an apomorphy. This is an evolutionary novelty for the group under consideration; hence it is a concept which has a meaning only in a particular context. These are evidences for the existence of a group. Put another way, attributes shared in common are taken to indicate a shared evolutionary history.

A novel evolutionary trait that is unique to a particular species and all its descendents and which can be used as a defining character for a species or group in phylogenetic terms. An apomorphy that is restricted to a single species is termed an autapomorphy. It alone cannot provide any information about the phylogenetic relations of that species, although it can indicate the degree of divergence of a species from its nearest relatives. An apomorphy that is shared by two or more species or groups is termed a synapomorphy. Such traits define the strictly monophyletic groups or clades, which are the basis of cladistic classification systems.

Synapomrophy: The possession of apomorphic features by two or more taxa in common (*i.e.*, the features are shared, derived). If the two groups share a character state that is not the primitive one, it is plausible that they are related evolutionarily, and only synapomorphic character states can be used as evidence that taxa are related. Phylogenetic trees are built by discovering groups united by synapomorphies.

A character which is derived and because it is shared by the taxa under consideration, is used to infer common ancestry.

Plesiomorphy

An ancestral or primitive character. Features shared more widely than in a group of interest are pleiosomorphies. These are primitive for the group in question and cannot provide evidence for the group.

An evolutionary trait that is homologous within a particular group or organisms but is not unique to members of that group (compare apomorphy) and therefore cannot be used as a diagnostic or defining character for the group.

Synplesiomorphy

The possession of a character state that is primitive (pleisomorphic) and shared between two or more taxa. Shared possession of a synplesiomorph character state is not evidence that the taxa in question are related.

Discussion

Note that apomorphy and plesiomorphy are relative concepts. Their status depends on their position in the phylogeny. A character is an apomorphy at one branch of the tree, but is a plesiomorphy relative to all the branches after that.

Clades or species relate to each other in different ways:

- ☆ A species or clade is basal to another clade if it holds more plesiomorphic characters than that other clade.
- ☆ A clade or species located within a clade is said to be nested within that *clade.*

History of Cladistics

Hennig's major book, even the 1979 version, does not contain the term cladistics in the index. He referred to his own approach as phylogentetic systematics, as implied by the book's title. A review paper by Dupius observes that the term clade was introduced in 1958 by Julian Huxley, cladistic by Cain and Harrison in 1960, and cladist (for an adherent of Hennig's school) by Mayr in 1965.

From the time of Hennig's original formulation until the end of the 1980s cladistics remained a minority approach to classification. However in the 1990s it rapidly became the dominant method of classification in evolutionary biology. Cheap but increasingly powerful personal computers made it possible to process large quantities of data about organisms and their characteristics. At about the same time the development of effective Polymerase Chain Reaction (PCR) techniques made it possible to apply cladistic methods of analysis to biochemical and molecular genetic features of organisms as well as to anatomical ones.

Cladistics as a Successor to Phenetics

For some decades in the mid to late twentieth century, a commonly used methodology was phenetics ('numerical taxonomy'). This can be seen as a predecessor to some methods of today's cladistics (namely distance matrix methods like neighbour-joining), but made no attempt to resolve phylogeny, only similarities.

Cladograms

The starting point of cladistic analysis is a group of species and molecular, morphological, or other data characterizing those species. The end result is a tree-lie relationship diagram called cladogram or sometimes a dendrogram Greek for "tree drawing). The cladogram graphically represents a hypothetical evolutionary process. Cladograms are subject to revision as additional data become available.

Synonyms: The terms evolutionary tree, and sometimes phylogenetic tree are often used synonymously with cladogram, but others treat phylogenetic tree as a broader term that includes trees generated with a nonevolutionary emphasis.

Subtrees are clades: In cladograms, all organisms lie at the leaves. The two taxa on either side of a split are called sister taxa or sister groups. Each subtree, whether it contains only two or a hundred thousand item, is called a *clade*.

2-way versus 3-way forks: Many cladist require that all forks in a cladogram be 2-way forks. Some cladograms include 3-way or 4-way forks when there are insufficient data to resolve the forking to a higher level of detail. See phylogenetic tree for more information about forking choices in trees.

Chapter 11
Species Concept

In the history of biology, two names are most intimately connected with the problems of species definition, namely Linnaeus and Darwin. The former believed in the constancy and sharp delimitation of species while the latter believed in variation and overlapping or ever-changing pattern of species. But the views of both underwent change during the life of each. Linnaeus became less and less dogmatic with statements on constancy of species, while Darwin found it impossible to delimit species with his idea of evolution. He finally regarded species as something purely arbitrary and subjective. He concluded "In determining whether a form should be ranked as a species or a variety, the opinion of naturalists having sound judgement and wide experience seems the only guide to follow" (Darwin, 1859).

Biologists subsequent to the publication of Origin of Species (1859) were clearly divided into two camps, namely the followers of Darwin and those of Linnaeus. The discovery of Mendelian laws resulted in an even more unrealistic species concept among experimentalists. None of these, however, studied species in nature as natural populations.

A study of natural populations, later became the prevailing preoccupation of naturalists and as a result a number of controversies arose. These may be summed up as those considering nature of species as either:

1. Subjective or objective,
2. Scientific versus purely practical,
3. Degree of differences versus degree of distinctness,
4. Consisting of individuals versus consisting of populations,
5. Only one kind of species versus many kinds of species and
6. Morphologically defined versus biologically defined.

Of these, three aspects are stressed in most modern discussions on species, that: (i) they are based on distinctness, rather than difference and therefore are to be defined biologically rather than morphologically; (ii) they consists of populations rather than unconnected individuals; (iii) they are more succinctly defined by isolation from nonconspecific populations than by the relation of conspecific individuals to each other. The crucial species criterion is thus not the fertility of individuals, but rather the reproductive isolation of populations.

Most of the early definitions of species regarded it as aggregates of individuals, unconnected except by descent. Depending on the choice of criteria, it leads to a variety of "species concepts" of "species definitions".

Classification of Species

Taxonomic (Morphological) Species Concept

The doctrine of fixity was challenged by Lamarck (1809) and finally Darwin (1859) who recognized continuous and discontinuous variation and developed his taxonomic species concept based on morphology, more appropriately known as the Morphological species concept. According to this concept the species is regarded as an assemblage of individuals with morphological features in common, and separable from other such assemblages by correlated morphological discontinuity in a number of features. The supporters of this view believe in the concept of continuous and discontinuous variations. The individuals of a species show continuous variation, share certain characters and show a distinct discontinuity with individuals belonging to another species, with respect to all or some of these characters.

Du Rietz (1930) modified the taxonomic species concept by also incorporating the role of geographic distribution of populations and developed the morpho-geographical species concept. The species was defined as the smallest population that is permanently separated from other populations by distinct discontinuity in a series of biotypes.

The populations recognized as distinct species and occurring in separate geographical areas are generally quite stable and remain so even when grown together. There are however, examples of a few species pairs, which are morphologically quite distinct, well adapted to respective climates, but when grown together they readily interbreed and form intermediate fertile hybrids, bridging the discontinuity gap between the species. Examples *Platanus orientalis* of the Mediterranean region and *P. occidentalis* of E. United States. Another well known pair is *Catalpa ovata* of Japan and China and *C. bignonioides* of America. Such pairs of species are known as vicarious species and the phenomenon as vicariance or vicarism.

Morphological and morpho-geographical types of taxonomic species have been widely accepted by taxonomists who even take into account the data from genetics, cytology, ecology etc. but firmly believe that species recognized must be delimited by morphological characters.

The taxonomic species concept has several advantages:

1. It is useful for general taxonomic purposes especially the field and herbarium identification of plants.

2. The concept is very widely applied and most species have been recognized using this concept.
3. The morphological and geographical features used in the application of this concept can be easily observed in populations.
4. Even experimental taxonomists who do not recognize this concept, apply this concept in cryptic form.
5. The greater majority of species recognized through this concept correspond to those established after experimental confirmation.

The concept, however, also has some inherent drawbacks:

1. It is highly subjective and different sets of characters are used in different groups of plants.
2. It requires much experience to practice this concept because only after considerable observation and experience can a taxonomist decide the characters which are reliable in a particular taxonomic group.
3. The concept does not take into account the genetic relationships between plants.

Biological Species Concept

This concept was first developed by Mayr (1942) and defined species as groups of actually or potentially interbreeding natural populations, which are reproductively isolated from other such groups. The words 'actually or potentially' being meaningless were subsequently dropped by Mayr (1969). Based on the same criteria Grant (1957) defined species as a community of cross fertilizing individuals linked together by bonds of mating and reproductively isolated from other species by barriers to mating. The recognition of biological species thus involves: (a) interbreeding among populations of the same species and (b) reproductive isolation between populations of different species. Valentine and Love (1958) pointed out that species could be defined in terms of gene exchange. If two populations are capable of exchanging genes freely either under natural or artificial conditions, the two are said to be conspecific (belonging to the same species). On the other hand, if the two populations are not capable of exchanging genes freely and are reproductively isolated, they should be considered specifically distinct. The concept has several advantages:

1. It is objective and the same criterion is used for all the groups of plants.
2. It has a scientific basis as the populations showing reproductive isolation don't intermix and the morphological differences are maintained even if the species grow in the same area.
3. The concept is based on the analysis of features and does not need experience to put it into practice.

The concept, first developed for animals, holds true because animals as a rule are sexually differentiated and polyploidy is very rare. When applying this concept to plants, however, a number of problems are encountered:

1. A good majority of plants show only vegetative reproduction, and hence the concept of reproductive isolation as such cannot be applied.
2. Reproductive isolation is commonly verified under experimental conditions, usually under cultivation. It may have no relevance for wild populations.
3. Genetic changes causing morphological differentiation and those causing reproductive barriers do not always go hand in hand. *Salvia mellifera* and *S. apiana* are morphologically distinct (two separate species according the taxonomic species concept) but not reproductively isolated (single species according to biological species concept). Such species are known as compilospecies. Contrary to this, *Gilia inconspicua* and *G. transmontana* are reproductively isolated (two separate species according to the biological species concept) but morphologically similar (single species according to the taxonomic species concept). Such species are known as sibling species.
4. Fertility-sterility is only of theoretical value in allopatric populations.
5. It is difficult and time consuming to carry out fertility-sterility tests.
6. Occurrence of reproductive barriers has no meaning in apomicts.
7. Necessary genetic and experimental data are available for only a very few species.

Genetic Species Concept

Another idea closely related to the biological species concept is the genetic species concept. This assumes that the biological factors of gene flow and reproductive isolation are operative, but that the way to define species is by a measure of the genetic differences or distance among populations or groups of populations. In effect, this is really the numerical phenetic species concept using a quantitative measure of genetic, rather than morphological (or other), distance as the yardstick. This has its obvious difficulties in the simple fact that we rarely know the real genetic differences between populations. Newer techniques of measuring at least part of the genome via allozyme electrophoresis (*e.g.*, Gottlieb 1977a, 1981a; Crawford, 1983) are most helpful here, and the genetic divergence based on allelic frequencies can be measured by various statistics such as Nei's (1972) genetic distance. The data are not yet avaible to indicate general levels of genetic divergence for each of the levels in the taxonomic hierarchy in plants and they may never be fully meaningful even when available. At the higher levels the approach may be severely limited by the likelihood of parallel point (and other) mutations yielding uninterpretable degrees of divergence. At the specific and intraspecific levels, however, they should prove to be most helpful. A modification of this genetic approach will be the direct measure of genetic distance from DNA sequences, which is now gaining momentum in plants particularly from chloroplast DNA studies. Even when complete DNA sequence data are available for entire genomes, however, which will be extremely helpful in taxonomic and phyletic studies, problems surely will arise in the interpretation of the data. Passive sites, feed back mechanisms, duplicated sites, and so on will all have to be kept in mind. That is, the sequency alone will not provide the whole story; it will be the sequence plus how developmental considerations and intramolecular events affect the final coding from the sequency that will tell the tale. This level of understanding will be a long time in coming.

Bock has offered a modification of the biological species concept that puts it somewhat intermediate to that of the genetic concept. He recommends changing the words "which are reproductively isolated from other such groups" to "which are genetically isolated in nature from other such groups". This is not a genetic distance concept, but rather an emphasis on genetic, rather than reproductive factors which are responsible in nature for keeping populational systems isolated. The viewpoints are similar, however.

Paleontological Spcies Concept

Paleontologists, working with fossil materials, cannot deal directly with species concepts based on gene flow and reproductive isolation. Their material is often fragmentary, rarely shows population varaion even at the morphological level and a few localities of a particular taxon are ordinarily known. While paleontologists can adhere philosophically to the biological species concept, in practice they must seek other means of definition. Further, they deal routinely with the time dimension in which species appear and later disappear in the fossil record, much different from the single-time reference afforded by extant taxa. Paleontologists, therefore, often speak of paleospecies (Simpson, 1961) or chronospecies (George, 1956) in which arbitrary time (and/or morphological) limits are used to delimit paleontological species (Cook, 1899). These are essentially slices of time that allow workers to communicate about the ordered fossil diversity. A collection of paleospecies in a monophyletic succession has been termed a gens (Vaughan, 1905). In practice, therefore, the paleospecies is usually a time-oriented morphospecies (Sylvester-Bradley, 1956). Sometimes distinct character state gaps occur between forms at different time zones, thus according a good place for making a species break. But if evolution in a particular lineage is gradual, and if sampling is good, then no clear breaks may be discernible.

Evolutionary Species Concept

This concept was developed by Meglitsch (1954), Simpson (1961) and Wiley (1978). Although maintaining that interbreeding among sexually reproducing individuals is an important component in species cohesion, this concept is compatible with a broad range of reproductive modes. Wiley (1978) defines: "an evolutionary species is a single lineage of ancestor-descendent populations which maintains its identity from other such lineages, and which has its own evolutionary tendencies and historical fate". Although the biological species concept (as well as the genetic species concept) is useful in many ways, it does not by definition refer to evolution directly.

It is the fact of evolution that has made genetical species separate and that keeps them from always being sharply, clearly separate. It is also evident that the genetical definition of species has evolutionary significance. Still it is striking that the definition does not actually involve any evolutionary criterion or say anything about evolution.

It would apply equally well, or in fact a great deal better, to species that did not evolve. Given the fact that the genetical definition of species is constant with evolution, its lack of any direct and overt evolutionary element certainly does not invalidate it. Nevertheless it is desirable also to have broader theoretical definition that relates the genetical species directly to the evolutionary processes that produce it (Simpson, 1961).

As a result of this perspective, the evolutionary species concept was advocated by Simpson (1951) to read specifically: "An evolutionary species is a lineage (an ancestral-descendent sequence of populations) evolving separately from others and with its own unitary evolutionary role and tendencies. This definition is useful to give a time perspective to neonatologists and a phyletic perspective to paleontologists (as opposed to a purely phenetic concept). Simpson (1961) emphasizes that this concept avoids the difficulties with determining actual or potential levels of interbreeding and gene flow, and it allows some degree of interspecific hybridization (so common plants), provided that it does'nt interface with the basic "evolutionary role" of each species. Determining what these "roles" are might be problematical, but Simpson (1961) suggests they are equivalent to niches taken broadly to mean the multidimensional relationship of a taxon to its environment rather than just its microgeographic situation. This point has been extended by Van Valen in a precise ecological definition: "A species is a lineage (or a closely related set of lineages) which occupies an adaptive zone minimally different from that of any other lineage in its range and which evolves separately from all lineages outside its range" (Simpson, 1976). He calls this the ecological species concept (cf. ecospecies, defined below).

Cladistic Species Concept

The evolutioanary species concept might be especially appealing to the cladists, who would search for a concept to relate to dichotomous branches on a cladogram (*i.e.*, their species). Hence, Wiley espoused the adoption of the evolutionary species concept with a few modifications: "A species is a single lineage of ancestral descendent populations of organisms which maintains its identity from other such lineages and which has its own evolutionary tendencies and historical fate. Although this definition is similar to Simpson's, several minor changes make it even more compatible with the cladistic viewpoint. The emphasis on "single" for the lineage more nearly equates this to a single branch on a cladogram. The use of "maintains its identity from other such lineages" rather than "evolves separately from others" opens the possibility of use of synapomorphies in detecting such lineages rather than the more general phrasing. And finally the stress on "historical fate" instead of "evolutionary role" (*i.e.*, ecological context) is a significant shift from an ecological viewpoint to a historical context resulting from apomorphic changes within a single branches of a cladogram. That is to say, Wiley's (1978) definition, although apparently emboidying only minor alterations from that of Simpson (1961), is really different in substantial ways – so much so that it seems best to call this the cladistic species concept. Bremer and Wanntorp (1979), two other cladists, also favour this concept. So do Donoghue (1985) and Mishler (1985), although they call it the phylogenetic species concept. Lovtrup (1979), still another cladist, takes issue with Wiley's cladistic species definition and proposes in a more radical way to abandon the use of any species concept as "detrimental" in cladistic (his "phylogenetic") classification. He admits that the species as a Linnean category is probably "necessary in practical taxonomic work), but he stresses simple recognition of terminal taxa of cladograms as the meaningdful units of diversity. Wiley (1980) rejects this as a largely artificial approach to the problem and emphasizes the need for an evolutionary view in which the temini of the branching points are cast as species resulting from the evolutionary process. Willis has gone even further and stated that "each species is an internally similar part of a phylogenetic tree". Wiley responded generally favourably to this suggestion, but regarded it "as a special case of the evolutionary species concept.

Biosystematic Species Concept

In addition to the principal types of species concepts in current use as discussed above, numerous other perspectives exist. It serves here to sketch some of these other concepts to indicate the breadth of viewpoints even beyond that already detailed. These additional species concepts reflect a desire to have units that more nearly reflect the diversity of reproductive relationships beyond the limitations allowed by the Linnean hierarchy. Most of these have not received wide usage, but some have become helpful informal descriptors in specific situations. It is clear, however, that these experimental categories will not replace the conventional categories of taxonomic hierarchy.

The experimental taxonomic studies of Turesson (1922, 1923), Clausen *et al.* (1939, 1941) and other led to special categories of taxa to express the variations encountered in their reciprocal transplant and hybridization studies (*e.g.*, Valentine, 1949). The most common ones are ecotype, ecospecies, and coenospecies (Cain, 1953; Grant, 19600. The ecotype refers to closely related but ecologically distinct populations that are largely interfertile. Ecospecies are similar but hybrids between them have reduced viability, and coenospecies are not interfertile, even artificially. The species aggregate is used to describe a complex of species that simply will not sort out well taxonomically for a variety of reasons, but in which there is hope of eventual resolution; the components of a species aggregate have sometimes been called microspecies (Davis and Heywood, 1963). Manton (1958) advocated the use of the concetpt to refer to morphologically poorly defined cytological or genetical groups. Grant's (1957) species group is similar to the species aggregate concept, as is Mayr's (1931, 1969) superspecies.

Numerous categories have been proposed to deal with the units resulting from biosystematic investigations in which much effort is placed on interpreting reproductive limits of taxa. The most extensive list is given in Camp and Gilly (1943) in which twelve kinds ar defined: homogenon, phenon, paragenon, dysploidion, euploidion, alloploidion, micton, rheogameon, cleistogameon, heterogameon, apogameon, and agameon. It serves no purpose to indicate here all the definitions of these terms, but two are given as examples. The homogenon is "a species which is genetically and morphologically homogeneous, all members being interfertile" and the heterogameon "a species made up of races which, if selfed, produce morphologically stable populations, but when crossed may produce several types of viable and fertile offspring". All of these concepts are based largely on morphological and interbreeding criteria. The apogameon and agameon apply to apomictic groups. Love (1962) agrees with this biosystematic approach but he did not in practice use all of the categories. This did, however, lead him to recognize cytotypes as distinct species because of reproductive barriers even without morphological divergence. The comparium and commiscuum of Danser (1929) are similar to the coenospecies, but more stress is placed on the ability to hybridize on geographic factors. The coenogamodeme and syngamodeme of Gilmour and Heslop-Harrison (1954) are equivalent to coenospecies and comparium, respectively (from Grant, 1957). The syngameon of Lotsy (1925, 1931) is approximately the same as a breeding population or in some cases equating to biological species. Grant redefined it as "the sum total of species or semispecies linked by frequent or occasional hybridization in nature; a hybridizing group of species; the most inclusive interbreeding population". This is similar to Van Valen's multispecies concept: "A set of broadly sympatric species that exchange genes in nature". The recognition that some plant species often hybridize freely with neighbouring taxa, especially the weedy relatives of cultivated crops, led Harlan and De Wet (1963) to propose the concept of compilospecies, which "is genetically aggressive plundering related species of their heredities and in some cases ... may completely assimilate a species, causing it to become extinct. The semispecies concept has been used in various ways to refer to an intermediate position between species and subspecies. Mayr (1940) regarded these as clear geographic segregates of a good species but so morphologically distinct as to be treated almost as distinct species. This viewpoint was followed by Valentine and Love (1958). Baum (1972) stressed reproductive criteria and viewed semispecies as on the way to becoming species. This is similar to Legendre's (1972): "a group of actually and potentially interbreeding populations, which are chromosomally somewhat distinct, but not effectively isolated from other such groups".

Crun (1985) says "A species cannot be fully defined, nor can it be intuitively sensed. Although subjectivity is involved in decision making, a species is only as good as the knowledge and insights used in its delimitation. Certainly methodologies help. So do good sense and good judgement based on meaningful experiences, and the more the better".

Semi-species

A group of organisms that are taxonomically intermediate between a race and a species, with reduced outbreeding and gene flow, *i.e.*, with incomplete reproductive isolating mechanisms. Semi-species are thought to represent advanced stages of speciation.

Successional Species

A species of tree that replaces a more shade-intolerant pioneer or other successional tree species in a forest. The seedlings of a successional species can grow in the shaded understorey and mature into the dominant species. The process of succession continues until the most shade-tolerant species (or climax species) is established. Species most often associated with early succession of a forest include lichen, moss, shrubs, wild grasses and climbers.

Cryptic Species

In biology, a cryptic species is a group of species which satisfy the biological definition of species, that is, they are reproductively isolated from each other, but their morphology is very similar (in some cases virtually identical).

The individual species within the complex can sometimes only be separated using non-morphological data, such as from DNA sequence analysis, bioacoustics, or through life history studies. They can, but need not be, parapatric, quite often are sympatric and sometimes allopatric.

Evidence from the identification of cryptic species has led some to conclude that current estimates of global species richness are too low.

A related concept is the superspecies. This is a group of atleast two more or less distinctive species with approximately parapatric distributions. Not all cryptic species complexes are superspecies and vice versa but many are.

Semi-cryptic Species

A quite different kind of taxonomic problem from that associated with hybridization is represented by Semi-cryptic species, so-called because their differences are marked in anatomical, chemical, cytological or genetical characters rather than morphologically. In some cases they represent incipient species, but in others they appear to be long established taxa which have not diverged markedly in gross morphology. They form the opposite end of the spectrum from many groups of Orchidaceae, in which genetically defined units can encompass an extremely wide and heterogeneous range of morphological species.

Because of their semi-cryptic nature, there is often considerable taxonomic argument concerning the correct status of the ultimate taxa; the species recognized by some taxonomists become relegated to subspecies or even varieties by others.

The two taxa *Anagallis foemina* and *A. arvensis* both have $2n = 40$ and are extremely similar phenetically, but they are intersterile.

Mechanism of Speciation

Speciation may be defined as the process by which one (or more) species give rise to another (or other) species (Ross, 1974). This basically not the concern of taxonomists. It is for evolutionists to investigate and unravel the process involved. In modern times, however, the concerns of the two groups are so interwined and entangled that one finds it impossible to separate evolution from systematics. Evolutionary taxonomy has its very roots in 'evolutionary relationships'. Moreover, without an understanding of the process of evolution and speciation, the very bases of the biological species concept, *viz.*, breeding relationships and reproductive isolation cannot be understood.

Allopatry

Allopatric and allopatry are terms from biogeography, referring to organisms whose ranges are entirely separate, so that they do not occur in any one place together.

If these organisms are closely related (*e.g.*, sister species), such a distribution is usually the result of allopatric speciation.

Allopatric speciation, also known as geographic speciation, is the phenomenon whereby biological populations are physically isolated by extrinsic barrier and evolve intrinsic (genetic) reproductive isolation, such that if the barrier should ever vanish, individuals of the populations can no longer interbreed. Evolutionary biologists agree that allopatry is common method by which new species arise. The word is derived from the ancient Greek *allos*, "other" + Greek *patra*, "fatherland". By contrast, the frequency of other types of speciation, such as sympatric speciation, parapatric speciation and heteropatric speciation, is debated.

Evolution of reproductive isolation is generally thought to be an incidental by-product of genetic divergence of other traits, particularly adaptive changes that evolve through natural selection in response to different environmental conditions in separate geographic areas. Ernst Mayr, an evolutionary biologist and famous proponent of allopatric speciation, hypothesized that adaptive genetic changes that accumulate between allopatric populations cause negative epistasis in hybrids, resulting in sterility or inviability.

Allopatric speciation may occur when a species is subdivided into two large populations (dichopatric or vicariant speciation) for example by plate tectonic or other geological events, or when a small number of individuals colonize a novel habitat on the periphery of a species' geographic range (peripatric speciation). Because natural selection is a powerful evolutionary force in large populations, adaptive evolution likely cause the genetic changes, that results in reproductive isolation in vicariant speciation. In paripatric speciation, however, the genetic changes that are thought to occur within the peripatric isolate are more controversial.

Proponents of peripatric speciation contend that small population size in the peripheral isolate (sometimes referred to as a "splinter population") allows genetic drift, which can be a more powerful force than natural selection in small populations, to demonstrate complex genotypes, allowing the creation of novel gene combinations. Both forms need not be mutually exclusive; in practice, passive isolation or fragmentation as well as active dispersal seem to play a role in many cases of speciation (see Wikipedia, the free Encyclopedia)

A famous example of allopatric speciation is that of Charles Darwin's Galapagos Finches.

Sympatric and allopatric are two terms commonly used in explaining related species and speciation. Two or more related species, the geographical distributions of which have a broad overlapping zone, are called sympatric (Smith, 1966). Several examples have been cited from both the animal (Grant, 1963) and plant kingdoms (Stebbins, 1950). When two or more species have widely separate and non-overlapping distributions, they are said to be 'allopatric'. Two species of the genus *Senecio, viz. S. canus* and *S. antennariifolius*, have such discontinuous distributions in America. The former is restricted to western USA, while the latter occurs only in the east (Ross, 1974).

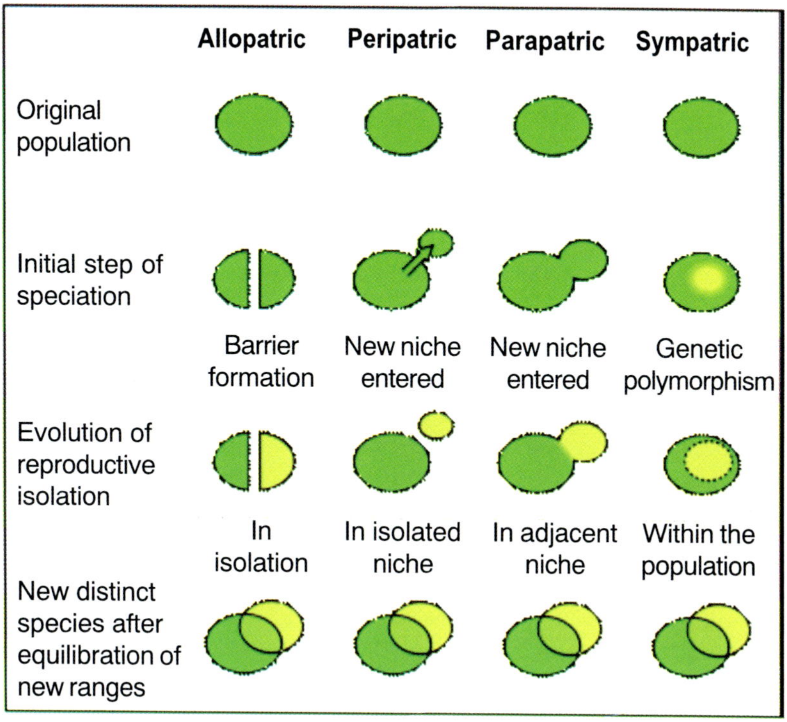

Figure 11.1

Sympatry

Sympatric speciation is the genetic divergence of various populations (from a single parent species) inhabiting the same geographic region, such that those populations become different species. Etymologically sympatry is derived from the roots sym- (meaning same, alike, similar, or fellow) and patry (meaning homeland or fatherland). In contrast to allopatry, populations undergoing sympatric speciation are not geographically isolated by, for example, a mountain or a river.

In multicellular eukaryotic organisms, sympatric speciation is thought to be an uncommon but plausible process by which genetic divergence (through reproductive isolation) of various populations from a single parent species and inhabiting the same geographic region leads to the creation of new species. In bacteria, however, the analogous process (defined as "the origin of new bacterial species that occupy definable ecological niches") is more common and occurs through horizontal gene transfer.

Debated since the beginning of popular evolutionary thought, sympatric speciation is still a highly contentious issue. By 1980 the theory was largely unfavourable given the void of empirical evidence available, and more critically the conditions scientists expect to be required. Ernst Mayr, one of the foremost thinkers in evolution, completely rejected sympatry outright, ushering in a climate of hostility towards the theory. Since the 1980s, a more progressive ideology hasbeen adopted. While still debatable, well documented empirical evidence now exists, and the development of sophisticated theories incorporating multilocus genetics have followed.

Parapatry

Parapatric and parapatry are terms from biogeography referring to organisms whose ranges do not significantly overlap but are immediatelyadjacent to each other; they only occur together in the narrow contact zone, if at all. Such organisms are usually closely related (*e.g.*, sister species), their distribution being the result of parapatric speciation.

Parapatric speciation is a form of speciation that occur due to variations inthe mating habits of a population within a continuous geographical area. In this model, the paerent species lives in a continuous habitat, in contrast with allopatric speciation and peripatric speciation where subpopulations become geographicallyisolated. Niches in this habitat can differ along an environmental gradient hampering gene flow, and thus creating a cline.

Parapatric, refers to species having slightly overlapping ranges of distribution. Ross (1974) attributes this to acute competition and cites the examples of certain American plant taxa which occur in serpentine soils while the related taxa occur in other types of soils as well. The former, though very formidable competitors in serpentine soils, cannot compete elsewhere. This creates a sharp borderline with a minimum of overlapping.

An example of this is the grass *Anthoxanthum*, which has been known to undergo parapatric speciation in such cases as mine contamination of an area. Thiscreates a selection pressure for tolerance to those metals. Floweringtime generally changes (in an attempt at character displacement – strong selection against interbreeding – as the hybrids are generally ill-suited to the environment) and often plants will become self pollinating.

Peripatry

Peripatric and peripatry are terms from biogeography, referring toorganisms whose ranges are closely adjacent but do not overlap, being separated by a natural barrier where these organisms do not occur – for example a wide river or a mountain range. Such organisms are usually closely related (*e.g.*, sister species), their distribution being the result of peripatric speciation.

Peripatric speciation is a form of speciation, the formationof new species through evolution. In this form, new species are formed in isolated peripheral populations; this is similar to allopatric speciation in that populationsare isolated and prevented from exchanging genes. However, peripatric speciation, unlike allopatric speciation, proposes that one of the populations is much smaller than the other.

Peripatric speciation was originally proposed by Ernst Mayr, and is related to the founder effect, because small living populations may undergo selection bottlenecks. Genetic drift is often proposed to play a significant role in peripatric speciation.

Chapter 12
Germplasm Conservation

Conservation can be classified into two major categories, *in situ* and *ex situ*. The latter can further be divided into five sub-categories

In situ Conservation

In situ conservation is an approach in which the species are conserved in their natural state through the establishment of nature or biosphere reserves, national parks, or by special legislation to protect endangered or threatened species. In this system, wild species and the associated natural ecosystems are preserved together to maintain the genetic integrity of the population and to allow natural evolution. The *in situ* conservation approach is not suitable for cultivated forms as there are no natural ecosystems to support them, although some workers had suggested that small areas could be preserved in which land races could be cultivated according to traditional methods to allow evolutionary process to continue. But, from the socio-economic point of view, this approach would generally not be acceptable. With ever-shrinking per capita availability of land, especially in South and South-East Asia (where arable land is most scarce), it would be difficult to set aside large areas for cultivation of low-yielding land races.

Ex Situ Conservation

Seed Genebanks

The most practical and the cheapest method of conservation of genetic resources of species producing orthodox seeds is through long-term cold storage of seeds. Orthodox seeds are those which tolerate a decrease in moisture content under low temperature. Such seed banks are also called genebanks. In case of those species which produce recalcitrant seeds, the seeds cannot be stored in gene banks, because recalcitrant seeds do not stand drying below a relatively high moisture content without a serious loss of viability, such as, in the case of cocoa, mango, durian and several other tropical fruits. Among species bearing orthodox seeds, there is genetic variation in storability; other things being equal, some species can be stored for longer periods than other. For instance, in leguminosae family, seeds of *Cassia*, *Trifolium* and *Lupinus* species could be cold-stored for over 100 years, whereas the seeds of groundnut have rather short cold-storability.

Depending on the duration of conservation, there are three types of conservation, namely, base collection, active collection, and working collection, as briefly described below.

Base collections: These comprise long-term stored materials which are disturbed only for purposes of regeneration. The materials conserved in base collections provide safety of genetic resources over generations. Base collections contain seeds with 5±1 per cent moisture content placed in sealed containers and stored at -18°C to 20°C.

Active collections: These comprise medium-term (10-15 years) storage and are used for the regeneration, multiplication, distribution, characterization and documentation. Seeds in active collections are stored at temperature around 0°C and moisture content of around 8 per cent.

Working collections: These comprise the genetic resources which are regularly used by the plant breeder, pathologist, entomologist, etc., who have no responsibility for maintaining them. However, even such materials also need not be grown every year and are required to be stored for three to five years. For such a storage, seed moisture content of about 8 to 10 per cent and temperatures of 5°C to 10°C is required.

During storage in a gene bank, it is necessary to control temperature and humidity to maximize the longevity of the seeds. Harrington (1963) indicated that seed longevity doubled for for each 5°C fall in temperature or for each 2 per cent drop in moisture content. This, he called, a "rule of thumb". This rule applied between 1°C and 50°C for temperature and between 5 and 14 per cent for seed moisture content. There is little interaction between the temperature regime and moisture content. In other words, the two rules based on temperature and moisture content apply independently.

Mass Reservoirs

Simmonds (1962) suggested the use of mass reservoirs or composite crosses of large number of diverse parents as a means of long-term conservation. Such reservoirs are subjected only to natural selection and thus remain under an evolutionary stage. It was suggested that, based on ecological, agromorphological and genetical similarities, different accessions could be bulked to constitute a mass population. Mass reservoirs, originating from intermating a set of parents, constitute a created genepool.

Field Genebanks

Many crop species, especially the perennial tree crops, produce recalcitrant seeds, and some of them even cannot be multiplied vegetativly. Hence, of necessity, such species are to be conserved as living field collections. Large scale *ex situ* conservation plots of vegetatively-propagated perennials have been or are in the process of being established in several countries. Malaysia has already put 500 hectares under oil palm germplasm. Indonesia has earmarked almost 1,000 hectares for conservation of coconut and other perennial crops conservation (Singh, 1982). Philippines is maintaining *ex situ* field collections of Southeast Asian germplasm of bananas. India has established a global collection of coconut germplasm in the Andaman and Nicobar Islands. Maintenance of such collections is rather costly and open to attack, by diseases and pests, thus it calls for careful management. On the other hand, it has the advantage of being continuously evaluated, and desired accessions can immediately be utilized directly or in breeding programmes. From the point of view of genetic conservation, field genebanks are, in most cases, totally unrepresentative of the range of genetic variability within the respective crop genepool and most of them do not constitute more than a fraction of the variability which sould conserved for the future.

Cold Storage and Freeze Preservation of Vegetative Parts

In the case of vegetatively-propagated crops, vegetative propagules such as rhizomes, corms, tubers, and cuttings are usually short-lived and deteriorate fast after collection, unless stored under appropriate conditions. Low temperature, (but not lower than 14°C) coupled with high humidity stores yam and sweet potato propagules rather satisfactorily for several months. Potato tubers can be stored adequately for five to seven months at 4-5°C and 90 per cent RH. But, these techniques do not free the curator from the job of raising the stocks every year or once in two years, which is not only costly but is also fraught with dangers of disease and pest attacks and many other hazards. Methods of medium and long-term storage such as cryo-preservation and *in vitro* conservation can be utilized for vegetatively propagated species.

In vitro Conservation

In vitro conservation is an important adjunct to usual conservations methods especially for those species which can not be conserved through orthodox seeds or whose seed production is very poor. Further, the tissue cultured materials are not exposed to pests and pathogens which usually appear on field collections. The materials can be stored at ultra low temperatures of liquid nitrogen (-196°C), thus giving high genetic stability and very long regeneration cycles. Finally, *in vitro* culture technique avoids the need of going through sexual cycle of about 10 to 20 years in several tree species. However this technique is applicable to only those species whose tissue culture and regeneration techniques are fully standardized. Some important vegetatively propagated species, *viz.*, mango, durian, and guava are yet not amenable to *in vitro* culture and would have to be maintained in field genebanks.

In vitro genebanks may also be grouped into two categories: (i) the *in vitro* base genebanks where cultures are maintained under conditions of cryopreservation; such genebanks are as yet non-existent, and (ii) the *in vitro* active genebanks wher cultures are maintained under slow growth, such as the existing collections of cassava at CIAT, potato at CIP and sweet potato at CIP, AVRDC and IITA. These genebanks are expected to be linked with breeder's working collections and field genebanks.

Conservation should ensure genetic fidelity of the material stored. The risk of genetic instability of *in vitro* stored material can range from minimal to considerable depending upon the culture systems used. Other things being similar, tissue culture systems derived from shoot, internodes or other aerial parts show greater instability as compared to those derived from root cultures.

Maintenance of Germplasm Resources

Over a number of decades many collections have been lost thus negating the scientific and financial outlay involved in their formation. Even large centres with modern facilities and adequate expertise have reported up to 50 per cent loss of samples from their original collections. The genetic fidelity of a sizable proportion of the older collections is often questioned. Further, most of the older collections contain an appreciable percentage of redundant duplicates.

Such problems can be the result of the incorrect choice of sample size for multiplication, of multiplication or regeneration in the wrong environments, lack of attention to the breeding system of the crop during multiplication and regeneration and numerous other factors. In fact, the methods of maintenance of material have not received the attention they deserve.

If germplasm is maintained correctly, the genetic constitution of the population at the time of its collection should be preserved. It is presumed that the original population was efficiently sampled and that the sample contains representative allelic frequencies and adaptive complexes. Neverthless, the maintenance of genetic integrity depends on how closely population genetics theory, biological, ecological and agricultural properties of the collections, the intended uses and the practicalities of multiplication are taken into account.

Characterization and Evaluation

Characterization means recording of state of those characters which are highly heritable, can be seen by the naked eye, and are expressed in all environments. Characterization is usually done by curators during multiplication of accessions.

Preliminary evaluation is made on a limited number of additional traits though desirable by a consensus of users of a particular crop. These traits can be assessed visually, but may not be expressed in all environments, for example reaction to disease and other stresses. Preliminary evaluation is often done by the curator by growing non-replicated observation rows.

Full evaluation is done by breeders in association with other allied disciplines and includes both quantitative and qualitative traits and those characters which are of interest to the breeder. This is usually based on multiplication multi-year replicated trials. For full evaluation, a multi-disciplinary team and an inter-disciplinary approach is needed, as is in vogue at International Rice Genebank at IRRI and in some of the national plant breeding programmes.

Documentation (Accession)

Documentation is an essential component of all germplasm resource activities and provides the linkage among curators, plant breeders and other researchers.

Systematic documentation of conserved materials and exchange of information stimulates germplasm exchange, evaluation as well as utilization. Hence an efficient system of data storage, retrieval and dissemination should be established.

Documentation could be manual or machine-based. For larger collections, such as the IRRI collection of rice where more than 70,000 accessions are held and for each accession about 45 descriptors (traits) are to be recorded, the volume and complexity of the data is very high and could not be handled manually. For such cases, suitable machine-based data management systems should be available.

Chapter 13
Practicals

Palynology

Acetolysis Method

This procedure should be done under a hood. Do not breathe fumes or get acetolysis mixture on skin or clothes.

1. Prepare acetylation reagent – 9 parts acetic anhydride : 1 part concentrated sulphuric acid. Add acid slowly to the anhydride. Prepare just enough for one day's use; it will not keep longer
2. Remove perianth parts of the flower and powder dried stamens through a fine mesh screen into a funnel placed in a centrifuge tube. Wet or fresh pollen must be dehydrated in glacial acetic acid first. Note: 1 to 6 samples may be run simultaneously in separate tubes.
3. Pour 5 yo 10 cc of the acetylation mixture over the polleniferous contents of each centrifuge tube. Wash pollen off the slides of the funnels with the above mixture also.
4. Place centrifuge tube in a water bath under a hood. Equip each tube with a glass striring rod and stir contents at regular intervals. Bring water bath to a boil and cook for 10 minutes. Remove stirring rods and let bath cool.
5. Centrifuge tubes at about 1500+ rpms for 1-2 minutes, then decant. All large debris should be decanted. REMEMBER: IF ACETOLYSIS MIXTURE IS ADDED TO WATER A VIOLENT REACTION WILL OCCUR. DECANT SLOWLY.
6. Wash pollen in tubes with glacial acetic acid, centrifuge and decant.
7. Add distilled water to the tubes, shake thoroughly, centrifuge and decant.

8. Add a few drops of equal parts glycerine and distilled water to the tubes. Allow to stand for ca. 10 minutes, centrifuge and decant.
9. Mount pollen grains. Use a bacterialogical loop to remove pollen from centrifuge tubes. The question of the most satisfactory mounting medium has not been resolved. The most widely used medium is glycerine gelly, although some workers use lactic acid or other media. Cover slips may be sealed using paraffin or left unsealed.

Anatomy

Epidermology

Leaf characters especially epidermal characters are very much useful in taxonomy. Epidermal peelings are useful in providing clear picture of stomatal complex and type of stomata. The peelings are scraped with a blade or the leaves are applied with a thin film of quick fix or nail polish. In case of dry material peelings are obtained by soaking or boiling in dilute potassium hydroxide or Sodium hydroxide solution. These peelings are mounted and observed under microscope (10x and 40 x). The number of stomata and epidermal cells are counted.

Stomatal Index

Stomatal index is calculated by using the following formula:

$$\text{Stomatal Index} = \frac{\text{Number of stomata}}{\text{Number of stomata} + \text{Number of epidermal cells}} \times 100$$

Stomatal Frequency

Stomatal frequency is calculated by using optic devises, namely ocular micrometer and stage micrometer.

$$\text{Microscopic field area} = \frac{\text{Stage micrometer divisions under 40 x}}{\text{Occular micrometer division}}$$

$R = d/2$

Unit area πr^2

$$\text{Stomatal frequency} = \frac{\text{Number of Stomata}}{\text{Unit area}}$$

Leaf Architecture

The venation of leaves contains cells which are more resistant to decay and maturation. Many of the other tissues of leaves are not so resistant like vascular cells. Thus leaf skeletons are prepared by differential distribution of leaf tissue as out lined below.

Leaves are boiled for 10-20 minutes in 20 per cent NaOH solution up to just boiling point until they turn brown and epidermal cells appeared to bubble up from venation.

The leaf skeletons are washed and dehydrated in an alcohol series 50 per cent ethanol, 75 per cent ethanol leaving the leaf for about 10-15 minutes in each solution.

Preparations are stained with 1 per cent safranin and they are placed in 95 per cent ethanol for about 20-30 minutes.

They are destained in absolute alcohol until desired leaves of stain was obtained for preparation of slides.

Attribution of Venation

1. *Venation type*: Several venation types are there. Pinnate venation is common in dicotyledones
2. *Venation subtype*: Mixed branched dromous, semicosmidodromous, reticulatedromous
3. *Number of secondaries*. Veins on both sides of the midrib.
4. *Primary vein size*: Determined midrib between the leaf apex and based on the ration of vein width to the leaf width.

$$\text{Primary vein size} = \frac{\text{Vein width at the base}}{\text{Leaf width}}$$

5. *Primary vein course*: Straight, markedly curved, sinuous, zigzag, smooth, changes in the direction of the curvature and bend.
6. *Number of areoles*: Per square mm. The smallest areoles of leaf tissue, surrounded by veins is called areoles.
7. Number of veinlets entering into areoles for microfield area:

Angle of divergence: Measured between the branch and continous of the source varying main above the point of branching.

Imperfect: Measure of irregular shape and size.

Areolation shape: a. Triangular, b. Quadrangular, c. Pentagonal, D, Polygonal, e,. Round, F. Irregular

Areolation size: Vary in size. more than 2 mm – very large

2-1 mm - large

1-0.3 mm – medium

0.3 – 0.1 mm – small

Secondary bifurcate/not.

Palmatel: Leaves with more than one primary veins.

Crystals present or absent

Intramarginal condition present/absent.

Disposition of veins absent/poor.

Angle of secondaries, gradually increase from base or not.

All the above attributes have to be worked out in the plants to be studied.

Phytochemistry

Phytochemical Methods

For qualitative detection the plant material – root, bark, leaves and seeds – are collected from wild plants and shade dried and powdered coarsely. 200 g of dry powder is extracted with petroleum ether at room temperature using Soxhlet apparatus till the liquid is clear. The extracts are then filtered and concentrated under vacuum. The material in the Soxhlet apparatus is air dried and then extracted with different solvents, *viz.*, Benzene, chloroform, ethyl acetate, Methanol and water in order of polarity and subsequently concentrated to get their corresponding residues.

Screening Tests for Secondary Metabolites

1. Detection of Alkaloids

The individual extracts are dissolved in chloroform. The solution is extracted with diluted H_2SO_4 or diluted HCl and acid layer was taken and tested for the presence of alkaloids.

a. *Mayer's Test*

To the acidic solution, Mayer's reagent (Potassium mercuric iodide solution) was added. Cream coloured precipitate indicates the presence of alkaloids.

b. *Wagner's Test*

To the acidic solution, Wagner's reagent (Iodine in potassium iodide) was added. Brown precipitate indicates the presence of alkaloids

c. *Hagner's Test*

To the acidic solution, Hagner's reagent (Potassium bismuth iodide) is added. Reddish brown precipitate indicates the presence of alkaloids

d. *Ammonium Reinckate Test*

To the acidic solution, ammonium reinckate solution was added. Flocculent pink precipitate indicates the presence of alkaloids.

2. Coumarins

One ml each of various extracts is treated with alcoholic sodium hydroxide. Dark yellow colour shows the presence of coumarins.

Ether, methanol and water extracts were tested for the presence of coumarins. The ethrel solution of the three extracts are evaporated and dissolved in water separately. UV fluorescence of the aqueous solution and the increase in intensity after the addition of 10 per cent NH_4OH indicates the coumarin presence.

3. Flavonoids

Five ml of each extract is separately dissolved in one ml each of alcohol and then subjected to the following tests.

a. *Ferric Chloride Test*

A few drops of neutral ferric chloride solution are added to one ml each of the above alcoholic solution. Formation of blackish red colour indicates the presenc of flavonoids.

b. *Shinoda's Test*

To one ml each of alcoholic extract, a small piece of magnesium ribbon or magnesium foil is added, and a few drops of conc. HCl are added; change in colour from red to pink shows the presence of flavonoids.

c. *Zinc-HCl Reduction Test*

A pinch of zinc dust and a few drops of concentrated HCl are added to alcoholic extract. Magenta color indicates the presence of flavonoids.

d. *Lead acetate test*

To one ml of alcoholic extract, a few drops of aqueous basic lead acetate solution were added. Reddish brown bulky precipitate indicates the presence of flavonoids. 5 ml of each extract is tested for the presence of different flavonoids and inferred by their colour reactions with different reagents.

Colour Reactions of Flavonoids with Different Reagents

Reagent 115 per cent NaCl	*Reagent 2 Conc. H_2SO_4*	*Reagent 3 Mg + HCl Hot*	*Reagent 4 Sodium Amalgum*	*Flavonoid Type*
Pale yellow	Pale yellow	No change in colour	No change in colour	Dihydrochalcones
Yellow	Intense yellow to red	Yellow to red	Red	Flavones
Yellow to brown by oxidation	Intense yellow	Red to magenta	Yellow to pale red	Flavonols
Yellow	Yellow	No change in colour	Red	Flavonones

4. Phenols

One ml each of the various extracts dissolved in alcohol or water was separately treated with a few ml of neutral ferric chloride solution. Any change in colour indicates the presence of phenols.

5. Anthraquinones

Fresh plant material is tested for the presence of anthraquinones. The fresh plant material is taken in a fresh test tube. Add 0.5 per cent Potassium hydroxide solution. This test tube is kept in a stand for 2 or 3 hours. Collect the extract. Take 1 ml of extract into a fresh test tube and add hydrogen peroxide, acetic acid and benzene. The mixture is treated with an equal amount of dilute ammonia. Appearance of red colour in the ammonia layer indicates the presence of carotenoids – anthraquinones.

6. Aucubins and Iridoids

Fresh plant material is tested for aucubins and iridoids. The plant material is chopped and treated with 5 ml of 1 per cent aqueous HCl. After 3-6 hours the extract is treated with 1 ml of Trim Hill reagent (10 ml of acetic acid, 1 ml of 0.2 per cent copper sulphate in water and 0.5 ml of concentrated HCl) and heated on water bath. The development of blue colour indicates the presence of aucubins (diterpenoids) while green and red colour indicates other iridoids and monoterpenoids.

7. Triterpenoids

The extracts are tested for the triterpenoids by Libermann-Burchard reaction. The extracts are dissolved in 0.5 ml of acetic anhydride followed by the addition of 0.5 ml of chloroform and 0.5 ml of concentrated HCl separately. Development of the red-violet colour indicates triterpenoids.

8. Steroids

Various extracts were dissolved in 5 ml of chloroform separately and subjected to the following tests.

a. Salkowski Test

One ml of concentrated sulphuric acid is added to the above solution and allowed to stand for 5 minutes after shaking. Lower layer turning into golden yellow colour indicates the presence of steroids.

b. Libermann Burchard Test

To one ml each of the chloroform treated extracts, a few drops of acetic anhydride, 1 ml of concentrated H_2SO_4 are added from the sides of the test tube and allowed to stand for 5 minutes. Formation of brown ring at the junction of two layers and the upper layer turning green indicates the presence of steroids.

c. Noller's Test

On ml each of the extract treated with chloroform is treated with a bit of tin foil and 0.5 ml of thionyl chloride, heated gently if necessary. Pink colour shows the presence of steroids.

9. Glycosides

5 ml each of various extracts are hydrolysed separately with 5 ml each of concentrated HCl and boiled for few hours on a water bath and hydrolysates were subjected to the following tests.

a. Legals Test

To 1 ml of each of hydrolysate 1 ml of pyridine and a few drops of sodium nitropruside solution are added and made alkaline with NaOH.

b. Borntrager's Test

To 1 ml each of hydrolysate, 1 ml of chloroform is added and the chloroform layer is separated. To this, an equal quantity of dilute NH_3 solution is added. Change in colour indicates the presence of glycosides.

10. Anthocyanins and Anthocyanidins

1 ml of each of methanol and water extract is tested for the presence of anthocyanins and anthocyanidins. Red colour in acidic aqueous solution of ethanol and water extract at pH 3-4 indicates the presence of anthocyanins and the change of colour with pH 8-9 indicates the presence of Anthocyanidins.

11. Saponins

One ml each of various extracts was separately mixed with 20 ml of distilled water and then agitated in a graduated cylinder for 15 minutes. Foam formation indicates the presence of saponins.

12. Volatile Oils

Ether extract was tested for the presence of volatile oils. 2 ml of the water extract was evaporated on a porcelain tile. Aromatic smell of the residue indicates the presence of volatile oils.

13. Tannins

Five ml each of various extracts was dissolved in minimum amount of water separately, filtered and the filtrate was then subjected to the following tests

a. *Ferric Chloride test*

To the filtrate a few drops of ferric chloride solution are added. A blackish precipitate indicates the presence of tannins.

b. *Gelatin test*

To the filtrate, gelatin (gelatin dissolves in warm water immediately) solution is added. Formation of white precipitate indicates the presence of tannins.

c. *Lead acetate test*

To the filtrate a few drops of aqueous basic lead acetate solution are added. Reddish brown bulky precipitate indicates the presence of tannins.

Table 1: Numerical Taxonomy of Capparaceae Members

Sl.No.	Attributes	*Cadaba fruticosa*	*Cleome aspera*	*Cleome felina*	*Cleome gynandra*	*Cleome viscosa*
1.	Habit: Herb/shrub/tree	Shrub	Herb	Herb	Herb	Herb
2.	Plant parts	Woody	Herbaceous	Herbaceous	Herbaceous	Herbaceous
3.	Stem	Erect	Prostrate	Prostrate	Prostrate	Erect
4.	Stem surface	Pubescent	Glabrous	Glabrous	Pubescent	Pubescent
5.	Leaf	Simple	Compound	Compound	Compound	Compound
6.	Number of leaflets	NC	3	3	5	3-5
7.	Leaf/leaflet shape	Elliptical	Linear	Obovate	Obovate	Obovate
8.	Leaf size L/B ratio	36 mm/12 mm	16 mm/4 mm	32 mm/4 mm	32 mm/4mm	32 mm/13 mm
9.	Petiole length	4 mm	2 mm	23 mm	40 mm	17 mm
10.	Spines (or) Prickles	Nil	Spines	Spines	Nil	Nil
11.	Position of flowers	Terminal	Axillary	Axillary	Terminal	Axillary
12.	Sepals size	38 mm	7 mm	31 mm	49 mm	22 mm
13.	Petals size	3 mm	0.53 mm	2 mm	4.5 mm	2 mm
14.	Petals colour	White	Yellow	Pink	White	Yellow
15.	Type of stamen	Polyandrous	Polyandrous	Polyandrous	Tetradynamous	Polyandrous
16.	Number of stamens	4	6	Numerous	6	12-24
17.	Stamens size	16 mm	3 mm	15 mm	17 mm	10 mm
18.	Flower size	38 mm	7 mm	31 mm	19 mm	22 mm
19.	Flower colour	White	Yellow	Pink	White	Yellow
20.	Stamen colour	Green	Yellow	Pink	Violet	Yellow
21.	Gynoecium length	18 mm 2 mm	4 mm 1 mm	10 mm 3 mm	10 mm 3 mm	8 mm 2 mm
22.	Fruit size	57 mm	44 mm	48 mm	106 mm	71 mm

Contd...

Table 1–*Contd...*

Sl.No.	*Attributes*	*Cadaba fruticosa*	*Cleome aspera*	*Cleome felina*	*Cleome gynandra*	*Cleome viscosa*
23.	Fruit diameter	13 mm	5 mm	12 mm	10 mm	11 mm
24.	Fruit texture	Pubescent	Glabrous	Glabrous	Pubescent	Pubescent
25.	Shape of fruit	Round	Flat like structure	Flat like structure	Cylindrical	Cylindrical
26.	Number of ovules	4	2	2	4	4
27.	Type of placentation	Axile	Parietal	Parietal	Parietal	Parietal
			Anatomical characters			
28.	Primary vein size	26 mm	10 mm	18 mm	20 mm	16 mm
29.	Number of secondary veins	8	–	4	14	6
30.	Secondary vein size	10 mm	–	5 mm	10 mm	7 mm
31.	Angles of divergence	60°	20°	45°	45°	53°
32.	Relative thickness veins	0.5 mm	0.5 mm	1 mm	1 mm	1 mm
33.	Inter secondary veins	Present	–	–	Present	Present
34.	Number of epidermal cells	Abaxial 140 Adaxial 72	160102	192176	192100	14072
35.	Number of stomata per unit area					
36.	Stomatal index	Abaxial 20 Adaxial	22.722	20.30	22.722	20
37.	Stomatal frequency	Abaxial 0.0210 Adaxial 0.0108	0.016 0.032	0.0288 0.264	0.032 0.016	0.0210 0.0108
38.	Multicellular hairs Macro/ Micro conical	–	–	Present	Present	–
39.	Unicellular hairs cylindrical	Present	–	–	–	Present
40.	Multicellular hairs cylindrical	–	–	–	Present	Present

Contd...

Table 1–*Contd...*

Sl.No.	*Attributes*	*Cadaba fruticosa*	*Cleome aspera*	*Cleome felina*	*Cleome gynandra*	*Cleome viscosa*
			Chemotaxonamy			
41.	Anthocyanins	–	+	–	–	+
42.	Aucubins	+	–	–	–	–
43.	Catecholic compounds	+	+	–	–	–
44.	Coumarins	+	–	–	–	+
45.	Flavonoids	–	–	+	+	+
46.	Dihydrochalcones	+	+	–	–	+
47.	Flavones	+	+	+	–	–
48.	Flavonoles	–	–	+	+	+
49.	Steriods	+	–	+	+	–
50.	Triterpenoids	+	+	+	+	–
51.	Saponins	+	+	+	+	+
52.	Phenols	+	+	+	+	+
53.	Volatile oils	–	–	+	–	–

Table 2: Conversion of Characters into Numerical Method

Sl.No.	Attributes	*Cadaba fruticosa*	*Cleome aspera*	*Cleome felina*	*Cleome gynandra*	*Cleome viscosa*
1	Habit	–	+	+	+	+
2.	Plant parts	–	+	+	+	+
3.	Stem	+	–	–	+	+
4.	Stem surface	+	–	–	+	+
5.	Leaf	–	+	+	+	+
6.	Number of leaflets	NC(-)	+	+	–	+
7.	Leaf shape	–	–	+	+	+
8.	Leaf size L/B ratio	+	–	+	–	+
9.	Petiole length	–	–	+	–	+
10.	Spiny	+	–	–	+	+
11.	Position of flower	–	+	+	–	+
12.	Flower size	–	+	+	+	+
13.	Flower colour	–	+	+	+	–
14.	Sepals size	–	+	+	+	+
15.	Petals size	+	–	–	+	+
16.	Petals colour	–	+	+	–	+
17.	Type of stamen	+	+	+	–	+
18.	Number of stamens	+	+	–	+	–
19.	Stamens size	+	–	+	+	–
20.	Stamen colour	+	+	+	–	–
21.	Gynoecium length	+	–	+	+	–
22.	Fruit size	+	–	–	+	+

Contd...

Table 2–*Contd...*

Sl.No.	*Attributes*	*Cadaba fruticosa*	*Cleome aspera*	*Cleome felina*	*Cleome gynandra*	*Cleome viscosa*
23.	Fruit diameter	+	–	+	+	+
24.	Fruit texture	+	–	–	+	+
25.	Shape of fruit	–	+	+	–	–
26.	Number of ovules	+	–	–	+	+
27.	Type of placentation	–	+	+	+	+
	Anatomical characters					
28.	Primary vein size	–	+	+	–	+
29.	Number of Secondary veins	+	–	+	+	+
30.	Secondary vein size	+	–	+	+	+
31.	Angle of divergense	+	–	–	–	+
32.	Relative thickness of veins	–	–	+	+	+
33.	Inter secondary veins	+	–	–	+	+
34.	Number of epidermal cells	+	–	+	–	+
35.	Number of stomata per unit area	+	–	+	–	+
36.	Stomatal index	+	–	+	–	+
37.	Stomatal frequency	–	+	+	+	–
38.	Multicellular hairs macro/micro conical	–	–	+	+	–
39.	Unicellular hairs cylindrical	+	–	–	–	+
40.	Multicellular hairs cylindrical hair	+	+	+	–	–
	Chemotaxonamy					
41.	Anthocyanins	–	+	–	–	+
42.	Aucubins	+	–	–	–	–

Contd...

Table 2–*Contd...*

Sl.No.	*Attributes*	*Cadaba fruticosa*	*Cleome aspera*	*Cleome felina*	*Cleome gynandra*	*Cleome viscosa*
43.	Catecholic compounds	+	+	–	–	–
44.	Coumarins	+	–	–	–	+
45.	Flavonoids	–	–	+	+	+
46.	Dihydro chalcones	+	+	–	–	+
47.	Flavones	+	+	+	–	+
48.	Flavonoles	–	–	+	+	+
49.	Steroids	+	–	+	+	–
50.	Triterpenoids	+	+	+	+	–
51.	Saponins	+	+	+	+	+
52.	Phenols	+	+	+	+	+
53.	Volatile oils	–	–	+	–	–

	1	2	3	4	5
1					
2	14				
3	25	66			
4	66	44	44		
5	48	40	59	62	

Morphology-Similarity Matrix

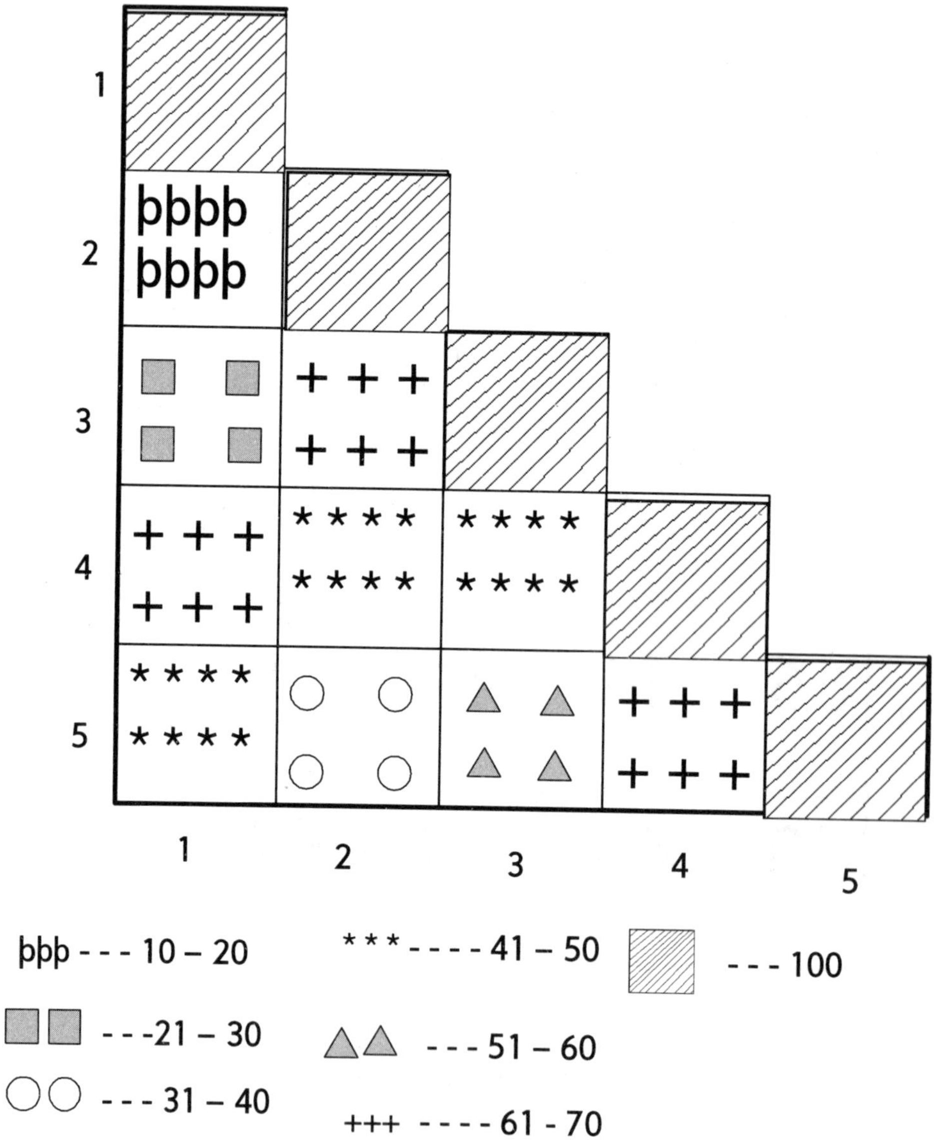

Morphology-Histogram

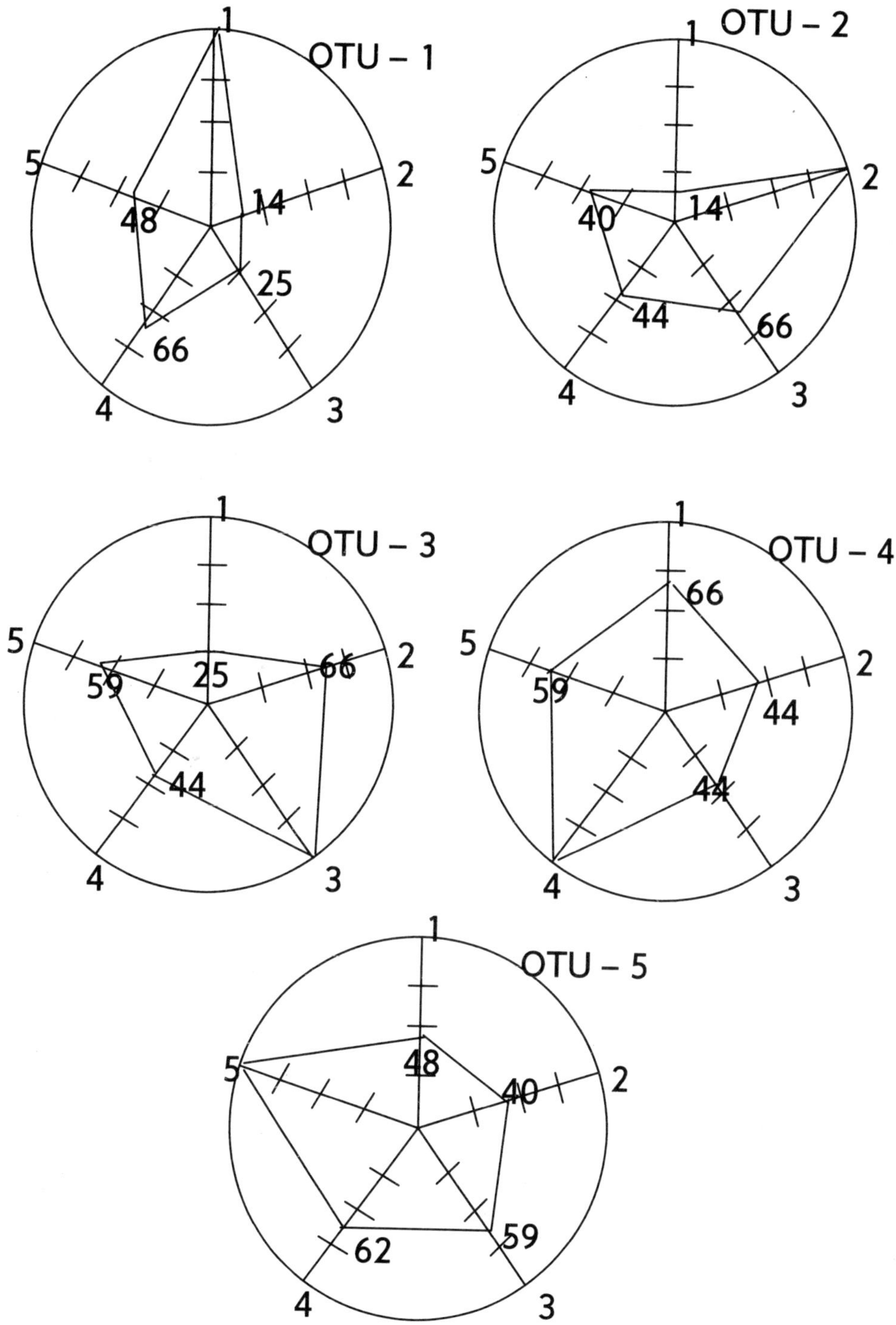

Morphology-Polygraphs

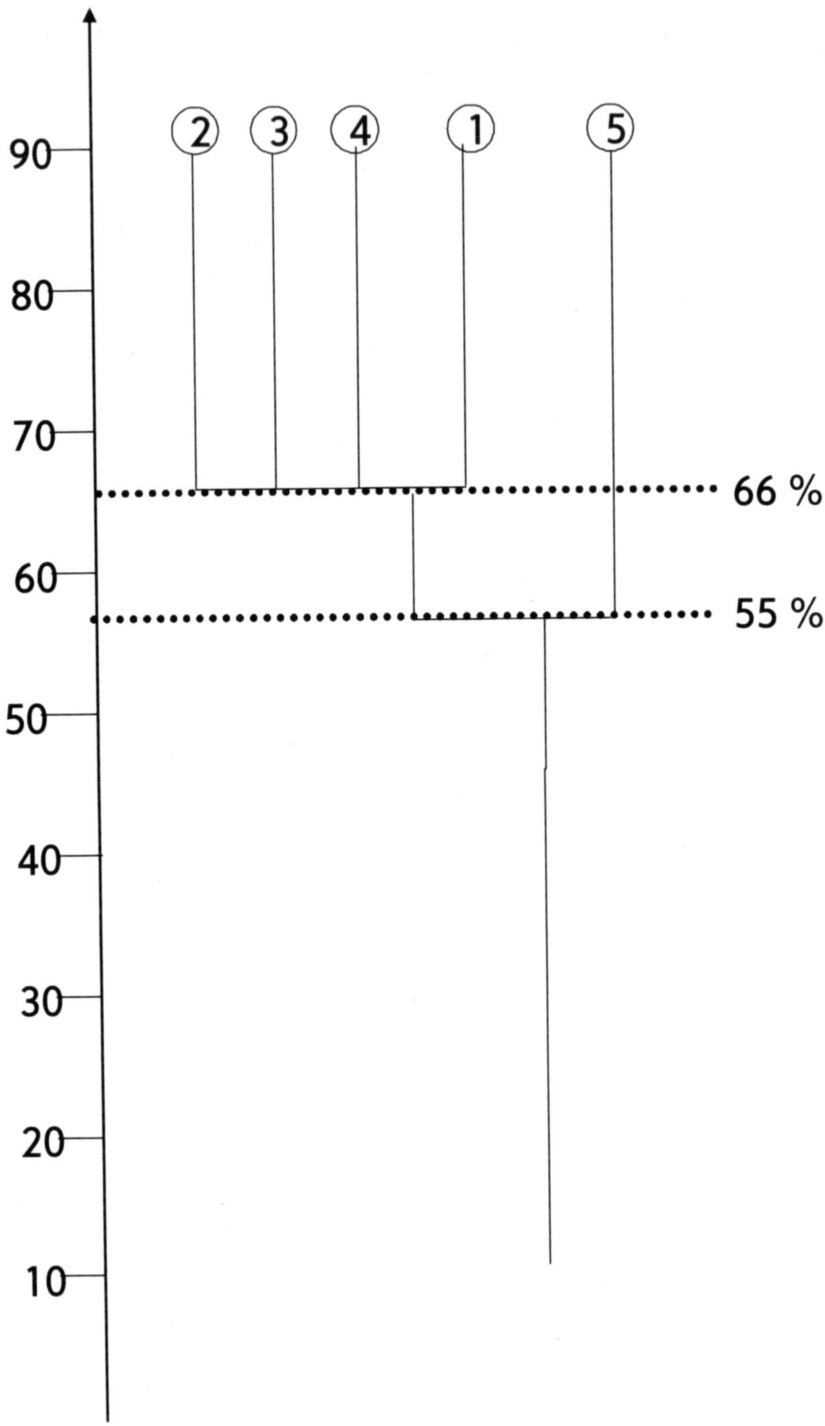

External Morphology - Phenogram

	1	2	3	4	5
1					
2	23				
3	46	46			
4	30	46	53		
5	76	69	53	38	

Anatomy-Similarity Matrix

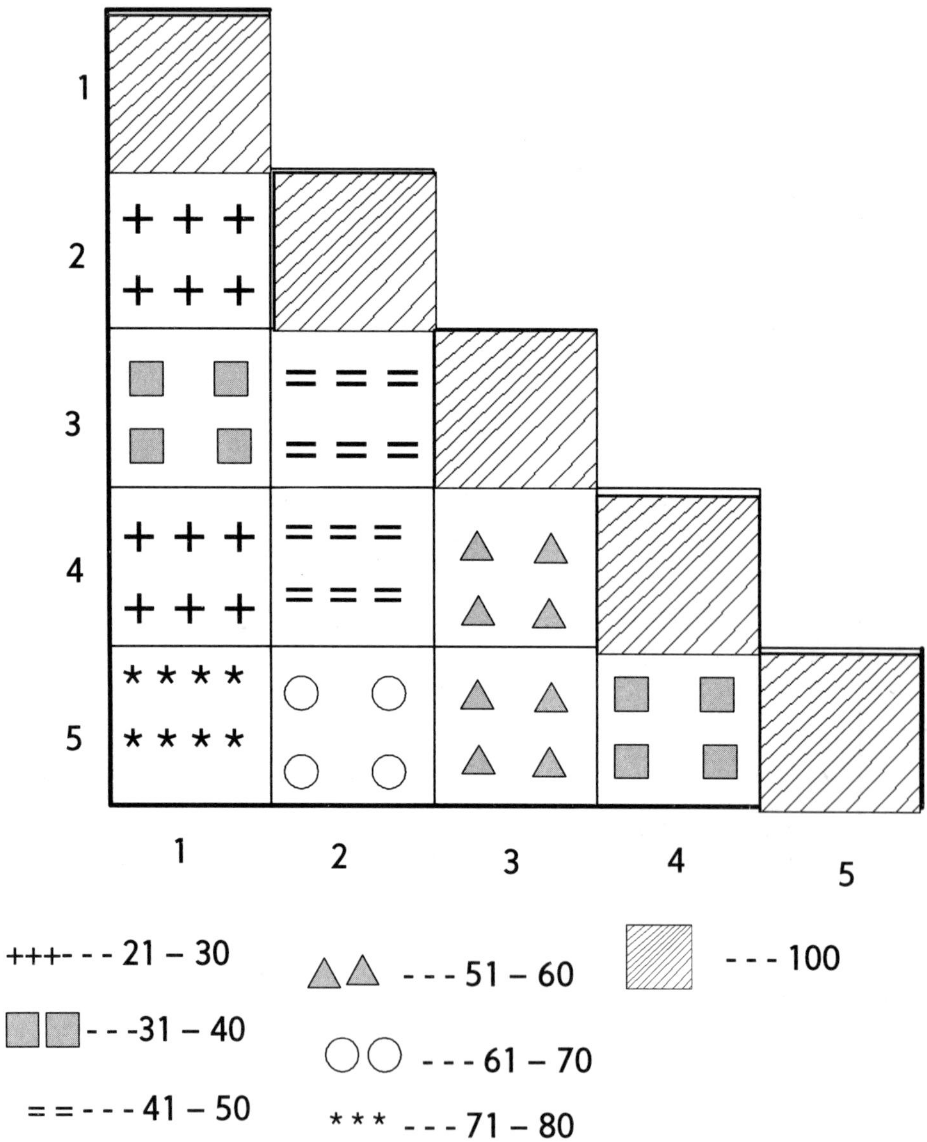

Anatomy-Histogram

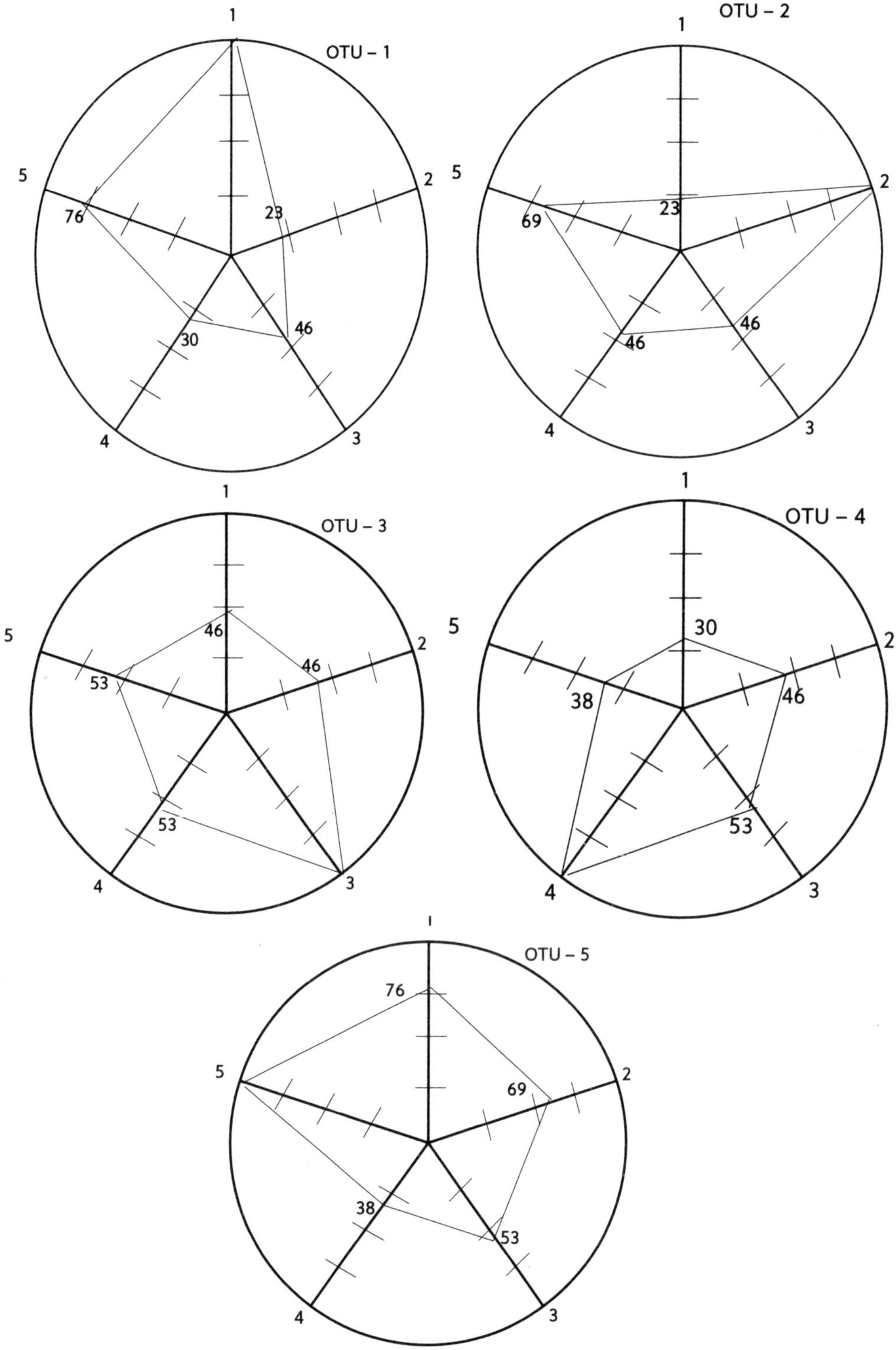

Anatomy-Polygraphs

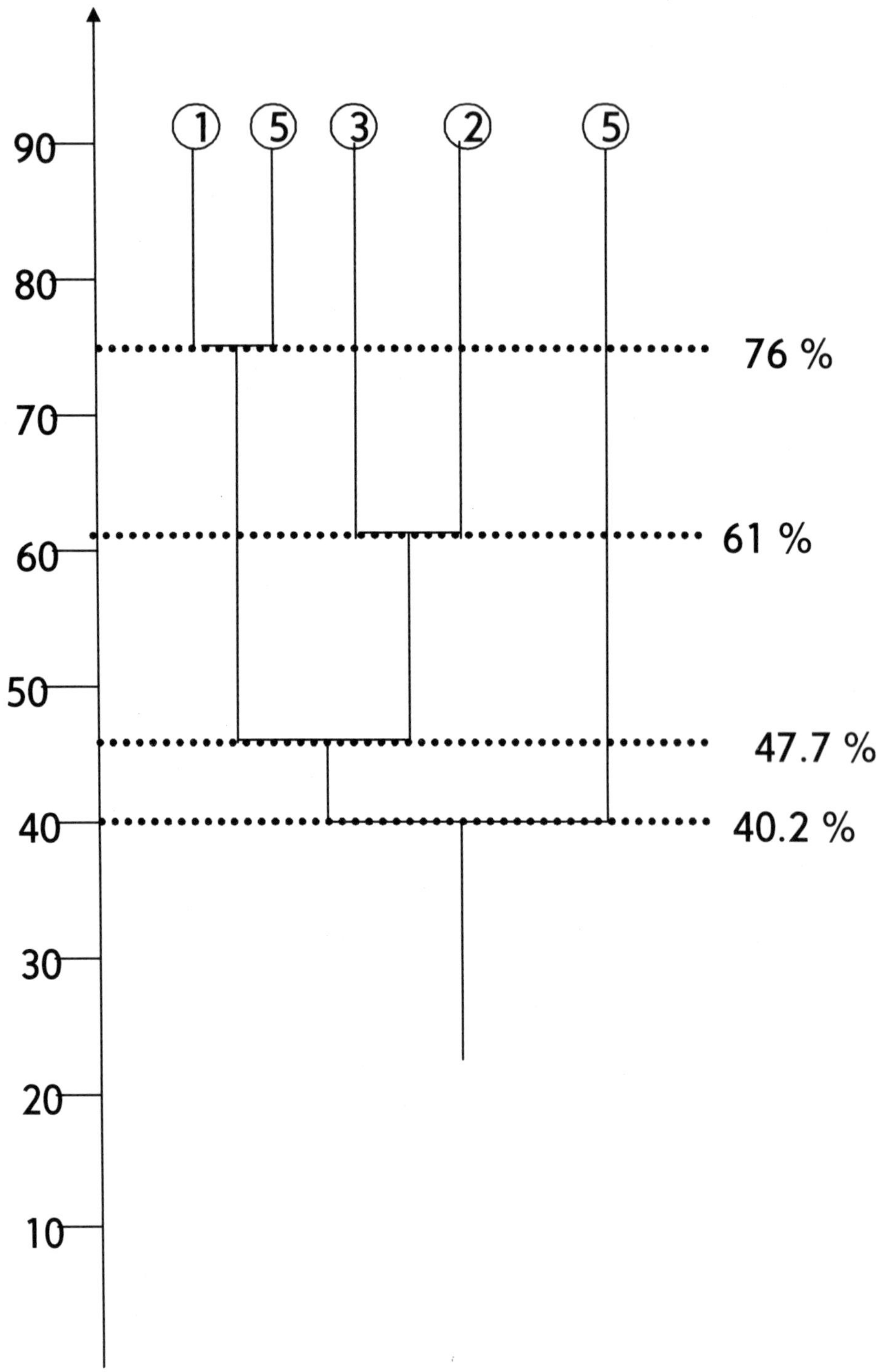

Anatomy - Phenogram

	1	2	3	4	5
1					
2	61				
3	46	46			
4	46	46	84		
5	53	61	69	46	

Chemotaxonomy-Similarity Matrix

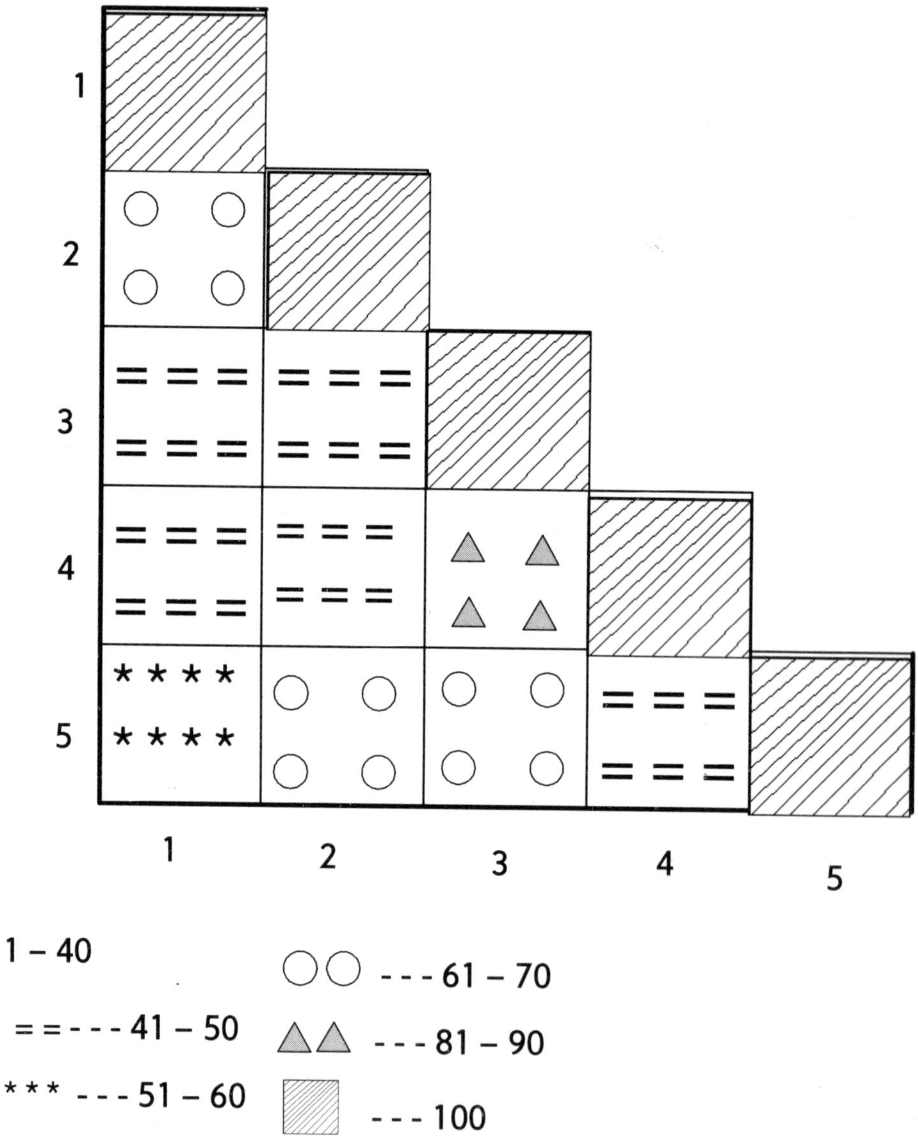

Chemotaxonomy-Histogram

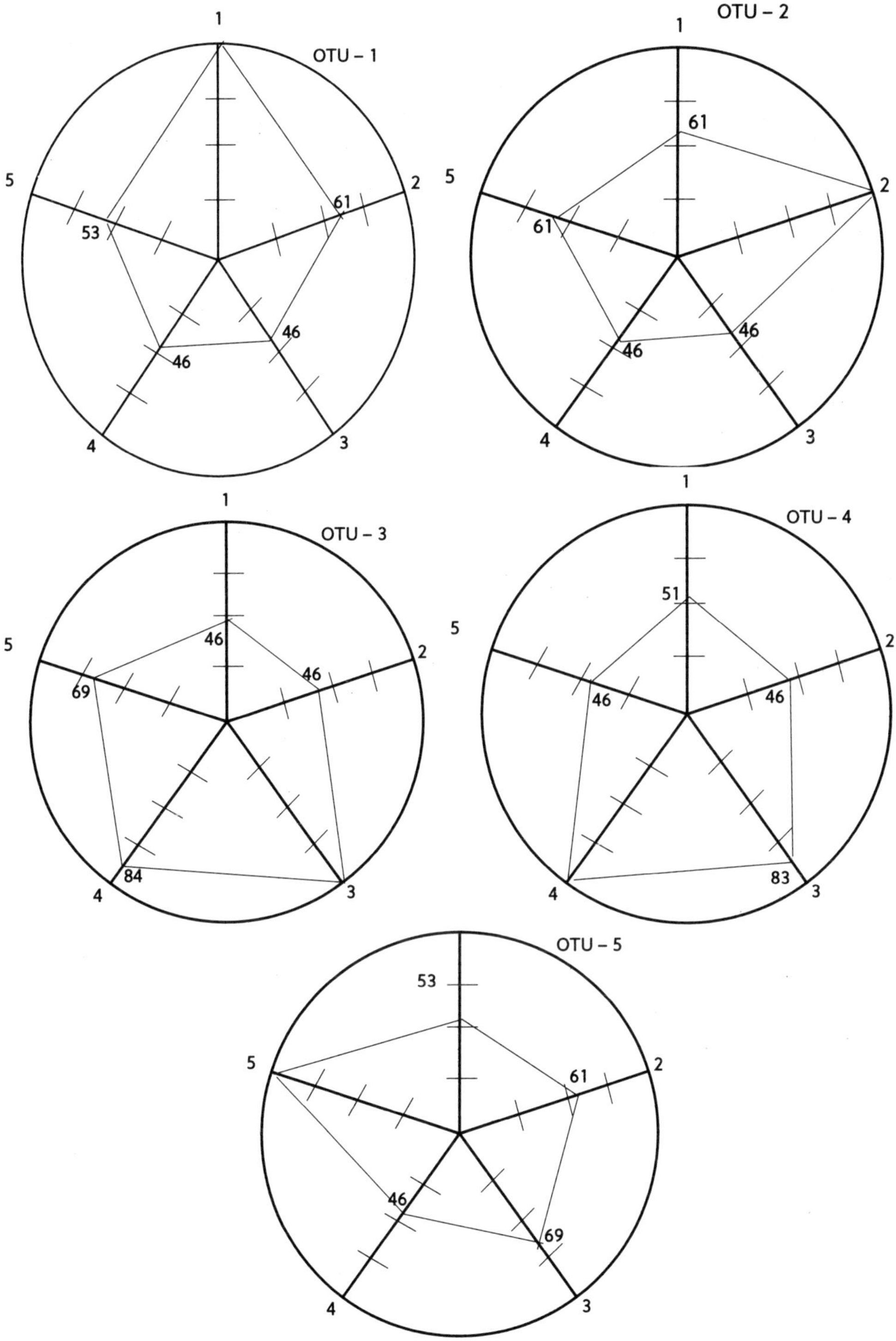

Chemotaxonomy-Polygraphs

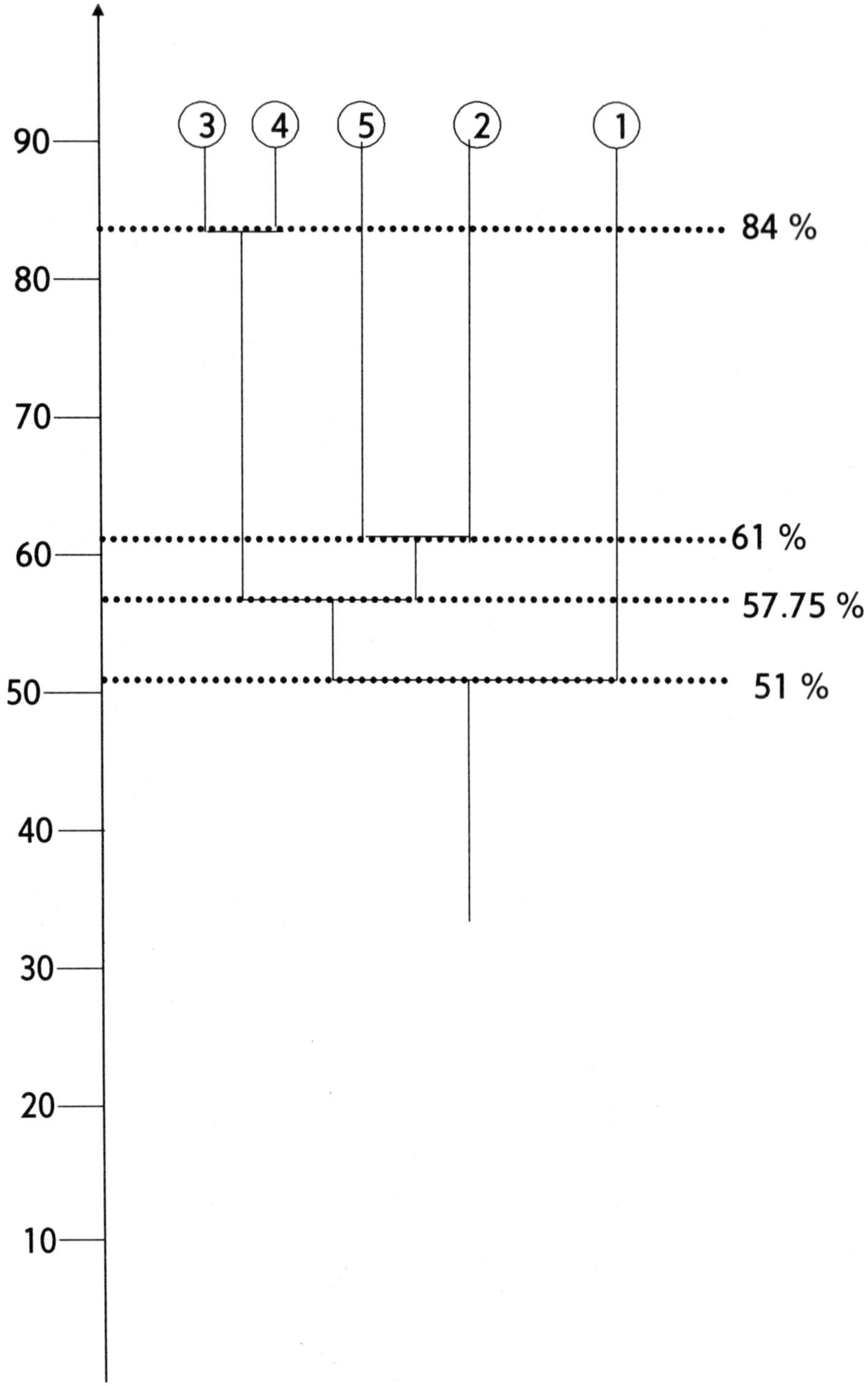

Chemotaxonomy - Phenogram

	1	2	3	4	5
1					
2	37				
3	28	47			
4	50	45	54		
5	52	50	50	52	

Total Similarity Matrix

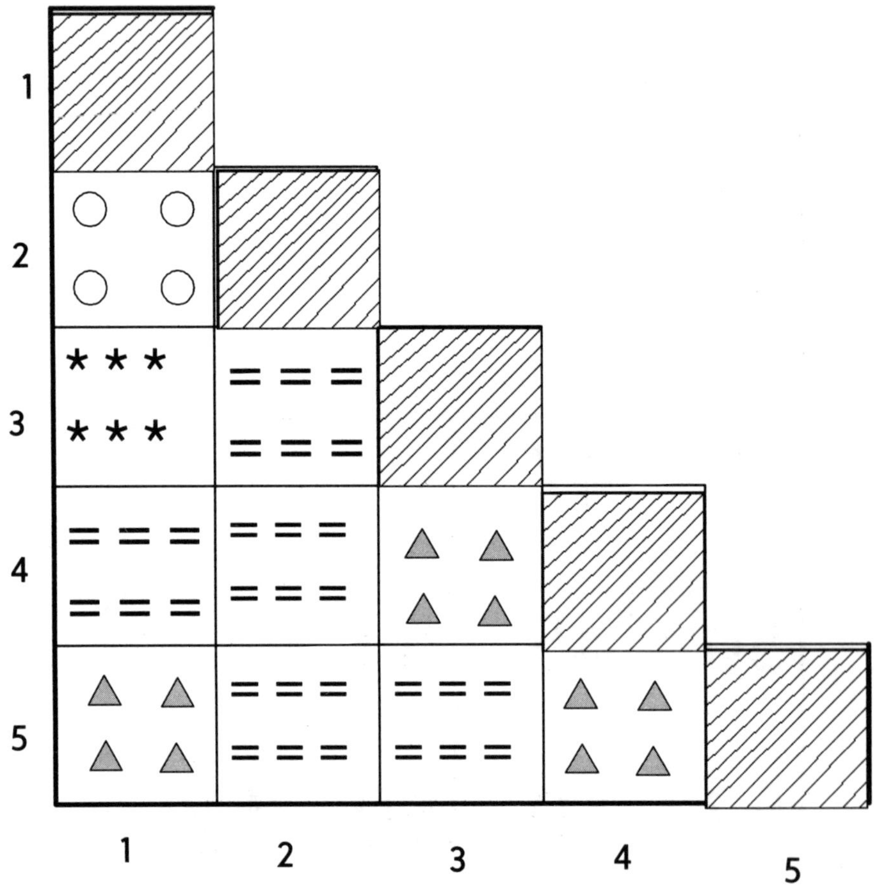

1 – 20

* * * - - - 21 – 30 = = - - - 41 – 50

○○ - - 31 – 40 ▲▲ - - - 51 – 60

▨ - - - 100

Total Histogram

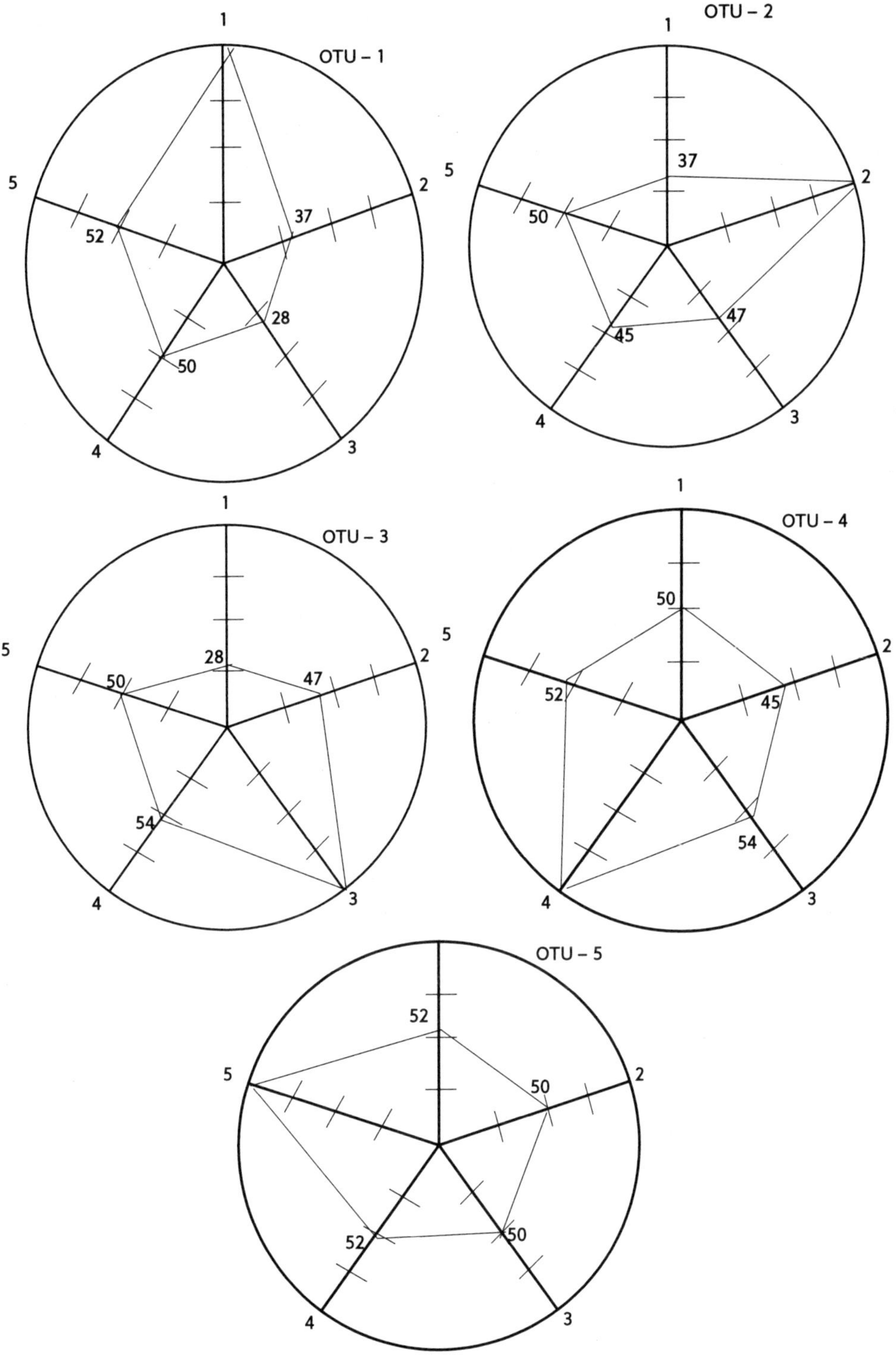

Total Polygraphs

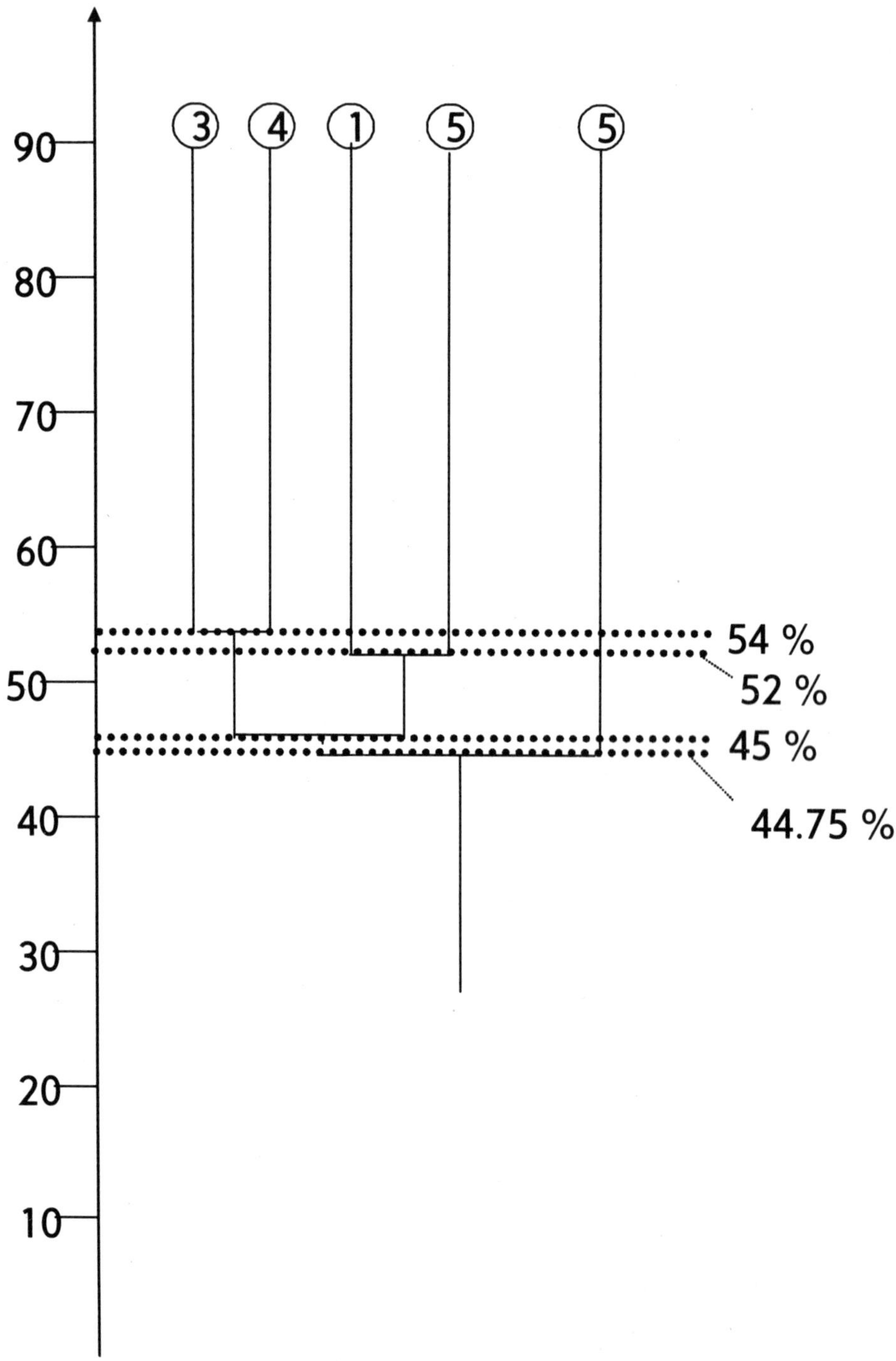

Total Phenogram

References

Davis, P.H. and Heywood, V.H. 1963. Principles of Angiosperm Taxonomy. Robert E. Kriger Publishing Company, Huntington, New York.

Hagen, J. B. 1984. Experimentalists and naturalists in Twentieth-century botany: Experimental taxonomy, 1920-1950. J. History of Biology 17: 249-270.

Heslop-Harrison, J. 1964. New Concepts in Flowering plant Taxonomy. Heinemann Educational Books Ltd.

Kubitzki, K. and Bayer, C. 2003. Flowering Plants. Dicotyledones.

Pullaiah, T. 2008. Taxonomy of Angiosperms. Regency Publications, New Delhi.

Radford, A.E. 1986. Fundamentals of Plant Biosystematics. Harper and Row Publishers.

Smocovitis, V. B. 1988. Botany and the evolutionary synthesis: the life and work of G. Ledyard Stebbins, Jr. Ph.D. thesis, Cornell University, Ithaca, New York.

Smocovitis, V. B. 1994. Organizing evolution: founding the Society for the Study of Evolution (1939-1950). J. History of Biology 27: 241-309.

Smocovitis, V. B. 1996. Unifying biology: the evolutionary synthesis and evolutionary biology. Princeton University Press.

Smocovitis, V. B. 1997. G. Ledyard Stebbins, Jr. and the evolutionary synthesis (1924–1950). American Journal of Botany 84: 1625-1637.

Solbrig, O.T. 1970. Principles and Methods of Plant Biosystematics. The MacMillan Company, London.

Stace, C.A. 1984. Plant Taxonomy and Biosystematics. Edward Arnold Publishing.

Stuessy, Tod F. 2009. Plant Taxonomy – The systematic evaluation of comparative data. Columbia Univ. Press, New York.

Venkata Ratnam, S. 2009. Plant Biosystematics. M.D.Publications, New Delhi.

Index